Elementary Equilibrium Chemistry of Carbon

Elementary Equilibrium Chemistry of Carbon

Grant Urry

Robinson Professor of Chemistry
Tufts University
Medford, Massachusetts

WILEY

A Wiley-Interscience Publication

JOHN WILEY & SONS

New York · Chichester · Brisbane · Toronto · Singapore

Library of Congress Cataloging in Publication Data:

Urry, Grant.
  Elementary equilibrium chemistry of carbon.

  "A Wiley-Interscience publication."
  Bibliography: p.
  Includes index.
  1. Carbon. 2. Carbon compounds. I. Title.

QD181.C1U77 1988   546'.681   88-27640
ISBN 0-471-84740-2

Printed in the United States of America

10 9 8 7 6 5 4 3 2 1

Preface

In the middle 1970s, driven by exigencies of the oil embargo, I was contemplating research directed toward developing a molecular photoelectrode. The boundaries of successful grantsmanship were readily discernible. The electrode should electrolyze water to hydrogen and oxygen. Ideally it should utilize coordination compounds of the transition elements in order to help justify the inordinate amount of government research money devoted to the earlier study of such compounds. The aim should be to achieve an efficiency that was some small fraction of the efficiency of natural photosynthesis.

The choice of hydrogen as a desirable goal was puzzling. Direct solar energy converters could have some practical purpose for use in stationary power plants. Hydrogen would seem to be preferred over these only for uses as a portable fuel. It seemed unlikely that hydrogen could ever enjoy an advantage over pure hydrocarbon fuels unless the fears over the greenhouse effect of increasing carbon dioxide content in the earth's atmosphere were to become dominant. The ability of natural photosynthesis to keep this problem in check has been demonstrated on several occasions in the past.

A photoelectrode capable of reducing carbon dioxide to compounds with a lower oxidation state for carbon seems to be an intellectual exercise of little practical utility; a gilding of the lily in light of the abundant inexpensive quantity of such reduction by natural photosynthesis. The great abundance and variety of biological species capable of this photoreduction makes it unlikely that solar energy will go to waste.

Pondering these matters, I recalled an experience my oldest child. She was engaged in a project of self amusement which my wife and I felt had a potential of damage to herself and/or her surroundings. In an effort to distract her from this presumably hazardous but long forgotten activity, her attention was directed to the television set. The inevitable acrobat was performing on the weekly Ed Sullivan Show. His *tour de force* was to balance on a ball, supported only by the tip of his index finger. My daughter glanced at the screen and said, "What's he doing that for?" and returned to more interesting pursuits.

The same question is pertinent here. Indeed, what were we doing this for? The honest answer to this question was not consoling. The main reason was to gain personal access to the funds generated by the then current enthusiasm of the government granting agencies. The implications of this answer were also disturbing. Over the years, driven by this motive, I had gradually allowed the same agencies to assume the major role in the direction of my research. Ironically, it was to seek freedom from such direction that I had chosen an academic career so long before.

Armed with this self-knowledge, the question became: "What course of research would be freely chosen in these circumstances?" It should address the compelling problem created by the world's heavy dependence on oil; it should have a sensible utility; and, it should draw from the great store of solar energy captured by photosynthesis.

The chemistry would necessarily have to utilize carbon in its elemental oxidation state and be capable of producing compounds of demonstrable current utility. It would necessarily require a return to a chemistry of an earlier time, a return to the chemistry that established at least two major chemical corporations, Union Carbide Corporation and American Cyanamid.

The reaction of elemental carbon with calcium carbonate in air can produce calcium carbide. Calcium carbide reacts with air under different conditions to produce calcium cyanamide. Both of these processes require impractically high temperatures.

A hypothesis, which since has proved to be embarrassingly naive, proposed that the high-temperature requirement was related to the high melting point of calcium oxide. In the context of this hypothesis, a lower melting basic oxide should promote similar chemistry. Potassium hydroxide suggested itself as a candidate for this role. Accordingly, in an informal preliminary experiment, a sample of carbon black was mixed with reagent-grade potassium hydroxide, the mixture was placed in a Pyrex vessel, and the vessel was evacuated. Heating the vessel with a gas-air torch caused the potassium hydroxide to melt. As soon as the carbon black came into contact with molten hydroxide, a rapid evolution of hydrogen gas occurred.

This fascinating result led naturally to the first experiment of many that have contributed to the current understanding of the chemistry presented in this research monograph.

A widely held popular misperception is that scientists search for the "truth". Some among the scientific fraternity, perhaps motivated by the ennobling ring of this description, happily enlist in this search. An undergraduate with an appropriate exposure to the Humanities will learn that this search is the chosen task of the philosopher. The natural philosophers direct this search for the "truth" of the natural, non-human, world. They use a version of the scientific method, by making observations, and interpreting these observations objectively as well as their understanding allows. Their interpretations are fitted into the body of existing understanding. The entire process is solely an intellectual effort to this point. There are inevitably defects in the fit of an interpretation into a body of understanding.

The scientist ponders these defects and *devises* further observations capable of furnishing information that might repair these defects. The interpretation of these *experiments* invariably exposes new defects in the body of understanding, and the iterative process continues. This cyclic process is more an exploration than progress.

Far from being a modern development, science is as old as humanity. Several excellent recent books describe scientific activities of humans through the ages. *The Discoverers* by Daniel J. Boorstin, *The Mapmakers* by John Noble Wilford, and *Revolution in Science* by I. Bernard Cohen come readily to mind in documenting scientific activity as a natural consequence of human nature.

Euclid at Alexandria, in the third century B.C.E., established the geometry that bears his name. The estimation of the circumference of the earth by his contemporary colleague at Alexandria, Eratosthenes, using the new Euclidean geometry, is as elegant a scientific experiment as any recent accomplishment. It is dramatic proof that the finest available scientific instrument is encased in bone above our necks.

The defects in our body of understanding furnish the scientist with all he or she needs to find happy endeavor. To an experimental scientist, happiness is in knowing what experimental task to perform next. I present this monograph with the confidence that the many defects in its body of understanding will provide many of you with numerous such occasions of happiness, as they have provided me in undertaking the work herein described.

It is a pleasure to acknowledge the kind assistance of many friends. Dr. William A. Doerner, of the Central Research and Development Department of E. I. duPont de Nemours & Company, recognized the potential implicit in the early results at Tufts University. He provided a former graduate colleague with high-pressure facilities. He also undertook the thankless task of directing the doctoral thesis of that student, Michael P. Santorsa. Doerner and Santorsa are due the credit for obtaining much of the qualitative information that established the main outlines of the high-pressure disproportionations. Walter Napolski of duPont performed the first quantitative analyses to prove that the high-pressure chemistry produced useful amounts of product in good yields.

Dr. J. Peter Jesson, while he was director of Physical Sciences of the Central Research Department at duPont, generously provided financial support and equipment for experimental work at Tufts University. O. R. van Buskirk designed and supervised the construction of the high-pressure microreactors, thus furnished. Dennis Babcock kept the equipment in good repair for the duration of the high-pressure studies and continues his kind assistance. Without the support of these duPont friends the experimental work that constitutes the major portion of Chapter IV would have been virtually impossible.

Another duPont colleague, Dr. Max Bechtold, after his formal retirement, continued to furnish friendly criticism and advice during the preparation of this monograph. His experience has made these contributions very valuable.

Graduate student colleagues at Tufts deserve the credit for almost all of the quantitative data contained in this monograph. Michael P. Crimmins carried out the bulk of the studies on the chemistry of graphite. Rueih Yuh Weng obtained the bulk of the data for the redox-disproportionations discussed in Chapter IV. Joseph P. Kakareka worked in both these areas, developing many of the techniques that made it possible to obtain quantitative data from the high pressure microreactors. He designed, built, and perfected the equipment required for the study of reactions in the solid state. He also explored the rudimentary tribochemistry of graphite, described here. It is in the nature of graduate students to assist their major professor. These gentlemen have gone far beyond the norm of kindness in assisting the author. Julie Lussier Cullen, while completing her thesis on a topic unrelated to this monograph, helped in many important ways.

Tufts colleagues, Professors Stephen Baxter, Barry Corden, Mary J. Shultz, David R. Walt, and recently Mark d'Alarcao provided a lively forum that was indispensable to the understanding of the work here reported.

Ted Hoffman, a senior editor at Wiley-Interscience, recognized the value of presenting this research monograph in a piece. He has given encouragement, genial advice, and much practical support during the long preparation of the manuscript. His intelligence and patience combine in the gentle excercise of his editorial responsibilities.

Lily has suffered the partial reading of drafts of this manuscript more often than should be required of a good wife. Throughout the period of the research reported here she has offered wise counsel and invaluable guidance innumerable times. Over many years she has cheerfully performed many normally shared domestic tasks. By this unselfish decision she has subsidized my research with that most important commodity, time.

Ultimately to the children goes an appreciation of their lively interest and critical discussions: Lisa, the emerging molecular biologist; Meggy, the practicing astrophysicist; Serena, the art conservator and proofreader *par excellence*; and Tony, through his voracious reading appetite, perhaps the most broadly educated of the group. They continue to make the author's whole life pleasant and meaningful.

Winchester, Massachusetts,

June, 1988

Contents

Chapter I

Introduction

🔲🔲🔲🔲🔲🔲🔲🔲🔲🔲🔲🔲🔲

COSMIC ABUNDANCE OF CARBON

The chemical importance of carbon is founded upon its cosmic abundance. Excluding helium as an element with minimal or no chemistry, carbon is third in abundance with only hydrogen and oxygen more abundant. Current estimates of cosmic abundance are given in Table 1.1. As stars cool, the loss of elements to interstellar space roughly parallels the relative abundances presented in the table. Interstellar space is therefore likely to be rich in hydrogen, oxygen, carbon, and nitrogen atoms as compared with second-row elements.

Table 1.1 Cosmic Atomic Abundances of Some Elements[a]

Element	Abundance	Element	Abundance
H	1.00	S	23×10^{-5}
He	0.09	Ar	6×10^{-6}
O	7×10^{-4}	Al	2×10^{-6}
C	3×10^{-4}	Ca	2×10^{-6}
N	9×10^{-5}	Ni	2×10^{-6}
Ne	8×10^{-5}	Na	2×10^{-6}
Fe	3×10^{-5}	Cr	7×10^{-7}
Si	3×10^{-5}	Cl	4×10^{-7}
Mg	3×10^{-5}	P	3×10^{-7}

[a]Current estimates relative to hydrogen[1]

Nitrogen and carbon are the only elements capable of an extensive structural chemistry among this group. Two orders of magnitude lower we find the next group of architecturally useful elements: calcium, aluminum, silicon, and phosphorus. This fact renders these elements insignificant among interstellar molecular species since they would probably be present in concentrations too low to be detectable by currently used spectroscopic methods.

There are only certain regions in interstellar space where molecular concentrations are detectable by these methods. Most of the current estimates come from observations on a few of these concentrations known as molecular clouds. Table 1.2 is the result of observations giving rise to the listing were made mainly by radio spectroscopy on the Taurus Molecular Cloud #1 (TMC-1).

Table 1.2 Interstellar Molecules (Dark Cloud TMC-1)[a]

Acetylenes and Polyacetylenes
$C \equiv C$
$C \equiv CH$
$HC \equiv CH$
$HC \equiv COH$
$H_3C - C \equiv CH$
$C \equiv C - CO$
$C \equiv C - CN$
$HC \equiv C - CN$
$C \equiv C - C \equiv CH$
$H_3C - C \equiv C - CN$
$HC \equiv C - C \equiv C - CN$
$HC \equiv C - C \equiv C - C \equiv C - CN$
$HC \equiv C - C \equiv C - C \equiv C - C \equiv C - CN$

[a]Detected by radioastronomy[2-7]

There are some surprises for a chemist in any listing of interstellar molecules. The cosmic abundance of iron might lead us to expect to see ferroorganic molecular species in these listings. Even likely simple groupings, FeC_6 for example, are not detected. The apparent absence of such iron-containing species could be an artifact of the observational method. Another surprise is the scarcity of molecules containing aromatic rings. This lack could have more significance chemically.

Table 1.2 excludes molecules containing fewer than two carbon atoms and hydrogenated or oxygenated to a point where they do not possess a high degree of chemical reactivity. While carbon monoxide, acetaldehyde, ethanol, and olefins could be included, they are less pertinent to our present discussions than the highly reactive acetylenes. Other molecular species such as CH, CH_2, and CH_3 are also relatively abundant and could comprise a source for cosmic diamonds to be discussed later in this chapter. The mechanism by which these interstellar molecules are formed is an intriguing puzzle.

Molecules formed in a diffuse vapor phase are constrained to dissipate the energy content of their components without the benefit of frequent intermolecular collisions. As a general rule, therefore, molecules arising from reactions of *hot* atoms, whether in space, in plasmas, or in arcs, usually are those that are metastable in condensed phases. As a logical corollary, these molecules are the most stable at equilibrium with radiant energy. Under these conditions it is not surprising that aromatic molecules are rarely found in interstellar space. With their intense absorption for radiant energy and with only intramolecular or radiation processes allowed, they would rearrange readily to linear molecules.

The polyacetylenes in Table 1.2 are such molecules. In the absence of moderate activities of water they should form a graphitic condensed phase. There is a threshold of water activity, however, above which acetylenes should form aldehydes; the oxides (carbohydrates) and nitriles (amino acids). The presence of amino acids in carbonaceous chondrites requires some such abiotic method of formations.[8]   The analysis of dust collected from the tail of Halley's comet also confirms the presence of many cyclic, unsaturated precursors to biologically useful compounds.[9] While the processes giving rise to comets is obscure, they represent yet another sample of interstellar condensed or accumulated matter. Carbyne phases are reported to be present in various carbonaceous chondrites[10] as possible carriers for the noble gases also found.[11]

Attempts[12] to reproduce conditions in earthbound laboratories for the study of such mechanisms are fraught with hazards. The state of carbon after evaporation from a hot star is most likely to be atomic carbon in a highly excited electronic state. Generation of hot carbon atoms in the laboratory by any currently available method leaves the question as to whether they are hot enough unanswered or equivocal. These hot atoms are probably quite diffuse in interstellar space. This produces a paradox in laboratory experiments: any method capable of producing sufficient quantities of hot carbon atoms for the detection and identification of molecular descendants is likely to produce them in a concentration much higher than that typical of interstellar space, since

the observational spectroscopic path lengths available in earthbound laboratories are infinitesimal compared with those in space.
This condition is also forced upon the experimentalist by the huge
difference in time scale for an experiment in an earthbound
laboratory as opposed to the interstellar experiment. The technique that should most closely approximate the interstellar experiment at the present time would be that in which diffuse plasma
beams are allowed to intersect. Analysis of products at different
scattering angles should be prolific of useful information. The
writer is unaware of any such experiments in progress.

Recent work using laser evaporation methods has given rise
to long linear polyacetylenes in a fashion proposed to be similar
to the interstellar processes.[13] In these experiments a molecular
beam of laser-evaporated polycarbon molecules was created by
expansion into a vacuum chamber swept by a carrier gas, mainly
He. A molecular beam was defined and created outside this
vacuum chamber. Various reactant gases were introduced to the
helium sweep, and the products in each case were analyzed by
time-of-flight mass spectrometry. Introduction of hydrogen gave
rise mainly to polyacetylenes, $C_{2n}H_2$ where $n = 4, 5, . . ., 10$.
When nitrogen was introduced to the sweep gas, dicyanopoly-
acetylenes, $NC\text{-}(C{\equiv}C)_n\text{-}CN$ where $n = 4, 5, . . ., 11$, were
produced. The analogous nitriles are found in interstellar space
where the likely nitrogen reactant is cyanyl radical.[14] In spite of
the difficulties previously described, this work has produced
interesting and stimulating results. The principal hypothesis, that
long linear carbon chains are produced in interstellar space by
particle bombardment of carbon grains, is not persuasive.

There are many chemists interested in hot-atom chemistry presenting work suggested as analogous to interstellar processes.[15,16] The reactants invoked, including atomic nitrogen, often are resident at low relative concentrations in interstellar space. Carbon stars currently appear to be the most probable source of interstellar carbon molecules. The atmospheres of such stars are, excluding the abundant atomic hydrogen, mainly poly-atomic molecules, such as NH, CH, CH_4, C_2, CN, HCN, CO, C_2H_2, HC_2, C_3, SiC_2, and HC_3.[17] There is very little water vapor in all of these stars, and only the hottest among them show the spectrum of atomic nitrogen.

Atomic silicon and silicon dicarbide are present in the atmosphere of carbon stars at a much higher atomic abundance relative to hydrogen than the abundance listed in Table I-A. The difference between abundances noted for carbon stars and those for molecular clouds possibly reflects molecular weight differences. The fraction of molecules achieving escape velocity from any gravitational body is proportional to this weight.

In that part of the earth available to scientific investigation, carbon is found in a relative abundance to hydrogen of 6×10^{-3}. This apparent discrepancy from estimated cosmic abundance could reflect the amount of carbon required to maintain the biosphere. The relative abundance of oxygen differs even more from cosmic abundance. These discrepancies illustrate the distortions produced by the examination of a severely limited sample, the earth's crust.

FORMS OF ELEMENTAL CARBON

Elemental carbon is found in several forms in the earth's crust. Graphite in its α form is the most abundant of crystalline carbon, its β form is the next, and diamond is the rarest form. A white carbon, reported to be a new hexagonal form with crystallographic spacings similar to a rhombohedral carbon, was found in the Ries Meteor Crater in Germany.[18] A controversy recently has developed over this definition, and the crystallographic data have been examined with the suggestion that the various forms of white carbon are silica or various silicates present as fortuitous impurities.[19,20]

This dispute arises between two competent and careful scientists. The use of powder x-ray diffraction for the complete characterization of new substances leaves considerable room for controversy. Unfortunately, the characterization of these supposedly novel materials as various known silica or silicate minerals used the same technique. It is proposed that these novel carbons are condensed phases of carbynes.[20] Pyrolysis of dust from certain carbonaceous chondrites yields volatile acetylenic molecules.[10] This fact certainly supports the notion of a carbynal phase in these meteorites. Additional support for this point of view is furnished by several careful studies where white crystalline condensation from carbon vapor, evaporated by several different procedures, all confirm such a hexagonal phase of carbon.

There is known terrestrial chemistry that might be invoked in this present controversy that could account for all of the observations cited.[21]   The reactions that occur when certain aromatic acetylenes are brought into proximity, as they would be in any interstellar condensation process, are macrocyclization reactions where transannular acetylenic bridges remain.   It takes little stretch of the imagination to suggest an analogous polymer that would be helical with crystallographic powder properties similar to quartz yet capable of generating substituted acetylenes upon pyrolysis.   Without such a stabilizing structure as these transannular macrocycles it is difficult to conceive of a crystalline phase wherein the highly reactive polyacetylenes lie side by side in any arrangement.   Other possible stable carbon networks in meteoritic material are those such as suggested by Hoffmann and colleagues.[22]

Fibrous carbon filaments also appear in such meteoritic fragments.[23]   A partial characterization of these has led to a suggestion that they arise from heterogeneously catalyzed decomposition of carbon-containing gases such as CO.   The presence of long polyacetylenes in interstellar clouds would seem to make such a source of carbon phases relatively unimportant.   Such a source would also not be likely to give rise to such a wealth of different crystallographic carbon phases.

Several novel polymorphs of carbon are produced by laser evaporation of carbon.[24]   One of these, a hexagonal form, may be the white carbon of Bavaria's Ries Crater.   It is apparently identical to the form produced by high-temperature evaporation.[18]   This hexagonal carbon exists in α and β forms as does hexagonal quartz.[25]

The rarity of diamond has created a fascination which was intensified for scientists when Lavoisier[26] demonstrated it to be a different form of carbon. The use of the term *allotrope* is avoided here since recent opinion[27] holds that the surface of diamond is covalently saturated with hydrogen. The prismatic edge of graphite apparently is substituted with mainly ketonic oxygens.[28,29] These facts would characterize graphite and diamond as different *compounds* of carbon, albeit of gigantic molecular dimensions. They are both nonstoichiometric compounds since composition is a function of crystal habit and size. The chemical characterizations of graphite will be discussed in the following chapter.

Diamond is found in meteoritic material,[30] and the question of cosmic diamonds in interstellar space should be considered. The molecular species deemed necessary for the epitaxial growth of diamond are relatively abundant in space.[27,31]

The evidence that meteoritic diamonds are formed in some collisional process of the parent meteor or a large meteoritic fragment is quite persuasive.[32] The evidence, however, is based upon x-ray powder crystallography. It remains a possibility that the smaller crystals arise from epitaxial growth upon a fortuitous solid particle in interstellar space. The solid particle should also serve as an efficient heat sink and passive radiator for the process.

In more recent work crystalline orientation in meteoritic diamond grains is used further to argue for anisotropic conversion processes, i.e. collisional processes.[33] Needless to say, only meteorites sufficiently large to survive passage through the earth's atmosphere are available for such studies. Epitaxial interstellar growth of the diamond grains is not excluded by the x-ray powder diffraction evidence. Most recently examination of the carbonaceous chondrites confirms the presence of a diamond phase[34] unlikely to have been formed by a collisional process. The authors also suggest that the estimates for interstellar graphite might be greatly exaggerated.

Diamond dust has been detected, with convincing characterization, in micrometeorites, collected in the upper atmosphere, by a combination of various energy dispersive spectroscopic methods with the analytical electron microscope and selective area diffraction.[35] These meteoritic grains contain unusually high carbon contents, some as high as 5% by weight suggesting some fractionation during the condensation process. The grains examined could not have been heated to temperatures very far in excess of $300°C$ and present no evidence of having been part of a larger meteoritic body. The idea that diamond *must* be formed at high pressures seems clearly to be wrong.[36] The origin of interstellar diamonds is still an open subject as well.[37]

ROLE OF CARBON IN THE ORIGIN OF LIFE

The possible biological importance of the interstellar polyacetylenes and their currently known behavior in condensed phases in oxidizing atmospheres[38] should be discussed. The notion of *panspermia*, assigning the origin of life to outer space, has recently been reinvoked.[39] In a fundamental sense, this notion should have some credibility if interstellar molecules, capable of generating those critical molecules consumed by early biological processes, are sufficiently abundant in space.   The "primordial soup" might then arise in the condensation process if sufficient water were present.   In later chapters the formation of these critical molecules, amino acids and carbohydrates, from other carbon sources typical of condensed phases will be discussed. With this chemistry the primordial soup might have developed in later stages of planetary evolution.   This suggestion is a liberal interpretation of the intent of the author of *Life Itself*[39] even if somewhat less spectacular.   In any event, the chemical processes necessary to the origin of live organisms had to occur someplace whether in space or on earth.

TECHNOLOGICAL USE OF CARBON

The use of carbon as a technologically important material is pre-historic.[40]   The recognition of the utility of fire allowed a wider exploration of geographically forbidding regions.   The construction of shelters with internal heating fires opened the temperate zones to year round dwelling.   The smoky wood fires and the natural discovery that charcoal, a serendipitous by-product of inefficient combustion, burned without the production of irritating smoke naturally would lead to widespread use of this form of elemental carbon.   It requires little imagination to see the extension of these circumstances to the earliest technologies of ceramics and metallurgy as arising from the higher temperatures attending charcoal fires.[35] It would be a remarkable coincidence if these technologies arose simultaneously in the third millennium B. C. E. without such a causal relevance.

Charcoal burning, where inefficient combustion conditions are intentional, probably produced the first reducing atmosphere with an attendant fortuitous chemical engineering. Reject iron[41] with a wide range of carbon content, from smelters used in producing the earliest bronze alloys, might well have furnished the experience that produced the age of iron.

The naiveté of this mythological account should not obscure the real probability of a fundamental role for elemental carbon in the technological ascent of the human species.

Corollary, secondary archeological evidence demonstrates a spread of these technologies from the less hospitable climates to the salubrious zones. In an unusual paradox, these charcoal based technologies arose in a tardy fashion in southeast Asia and China,[42] regions that in some respects were quite advanced compared with the source areas of charcoal technologies. The Chinese development of gunpowder in the first millennium apparently antedates such chemical applications of charcoal in the West and serves to dramatize this paradox. The appearance of metallurgy and ceramic technologies on the African subcontinent roughly coincides with their appearance in southeast Asia, as would be consistent with the technologically retarding effect of climate in regions where domestic fires were not used inside dwellings.

The chemistry of charcoal production is complex. It is produced in any wood fire unless air is supplied under forced draft, as in a forge. During the conversion of wood to charcoal, with limited air supply, many volatile constituents distill from the fire. Among these are wood tar, methanol, acetic acid, acetone, and long-chain aliphatic alcohols. The wood tar is a complex mixture that includes various phenols. Its historic use as an antimicrobial preservative arises from this phenol content. The increased use of wood for domestic heating has given many the experience of a mixture of the less volatile combustion products as creosote deposits that collect in chimneys. They are produced in considerable quantities in the popular air-tight stoves. Their presence in the chimney is often manifest when they ignite to produce destructive chimney fires. There are also gaseous by-products of charcoal burning such as hydrogen, methane, ethylene, acetylene, carbon monoxide and carbon dioxide. Much of the current production of charcoal burns off all of these useful by-products since other petroleum-based processes are more economical. Similar products are obtained when the wood is dehydrated by super-heated steam.[43] Most of these processes convert about 80% of available carbon to charcoal; the rest of the carbon is converted to combustion gases and volatile by-products.

Bone charcoal, still widely used as a decolorizing charcoal, is produced in processes similar to those for wood-charcoal production after the bones have been *degreased* and *degelatinized*.[44] Pyridine is one of the major volatile products of this pyrolysis.

There is an apparently natural tendency for the human intellect to attribute many deleterious effects to technological advancement. Ancient charcoal burning for technological uses is not immune from such criticism. Large areas of forest were denuded to produce charcoal. Such destruction of the forests in North Africa possibly contributed to the climatic changes that extended the Sahara Desert to the shores of the Mediterranean Sea. It is comforting to this writer to observe that the earth's atmosphere was able to overcome such devastation by making adjustments that removed that area from destructive human exploitation in a conclusive fashion.

The use of coal, coke, and oil in place of charcoal in more recent times has saved other forests from such destructive utilization. This more modern technology has allowed a population growth such that many are now concerned with the current production of carbon dioxide. They see catastrophe ahead, with the earth's atmosphere warming as a consequence of increased absorption of solar energy in the atmosphere. Such concerns are understandable, but they ignore the prevailing control that photosynthesis exerts over the carbon dioxide content of the troposphere.

More perceptive worriers are concerned that the massive clearances of tropical forests stimulated by population pressure pose a more serious threat to the regulation of carbon dioxide in the world's atmosphere. The concern could more profitably be directed to the goal of keeping the production of carbon dioxide from all sources, humans included, in balance with its solar reduction to carbon at its elemental oxidation state. The photosynthesis cycle is fed by solar energy of reasonably constant amount. Little that we can do will alter this state of affairs. If the amount of carbon dioxide increases, organisms that require a higher level of CO_2 activity will no doubt prosper and reduce the activity to the point where their growth is again restricted.) As the Sahara Desert proclaims, humans have room to make many mistakes. The solutions may not be desirable to most humans, but they could be less damaging than some well-intended efforts to maintain unchanged the current benefits we enjoy.

This monograph discusses a chemistry[45] that allows utilization of solar energy by participation in the carbon redox cycle. This process provides for the extraction of currently useful chemical compounds at the expense of forming carbon dioxide. In this respect it is like all other current chemical technologies. An abiding faith in the photosynthetically active biosphere's ability to deal with this and a knowledge that current means of producing these compounds generate relatively more carbon dioxide than the chemistry discussed here tells us that such chemistry could prove useful.

The formation of an effective cartel and the conditions giving this cartel control over the price of petroleum in 1973 shocked much of the industrial world into an unpleasant awareness. The dependence of the technologically advanced nations upon this particular fuel was dramatically illustrated. The artificially imposed crisis stimulated exploration for, conservation of, and more careful recovery of existing sources of petroleum. It also generated a frenzied search for alternate energy sources. The immediacy of this search was intensified by dire warnings that fossil fuels were being depleted at a rate that would cause their disappearance within the lifetimes of most of us.

The fear generated by these warnings was intended to influence our actions in many ways. If the "Chicken Little" syndrome is fostered, it must be assumed that either there are not persuasive logical reasons for a given action or that they are too weak to be viewed as effective a motivation as fear.

A popular notion that hydrogen is the fuel of the future is interesting to contemplate. The data in Table 1.1 certainly furnishes evidence for an infinitely available cosmic and terrestrial supply. Winning molecular hydrogen from the compounds in which it is found terrestrially still requires energy from some source. Electric power plants (nuclear or thermonuclear?), chemical reduction plants (using fossil fuels?), or solar conversion photoelectrodes all might be required to furnish a sufficient supply.

There are applications where hydrogen is uniquely applicable. In rocket engines where specific impulse and the related terminal-speed limitation require combustion products of the lowest possible molecular weight and internal heat content (c_p), it is the fuel of choice. Even in this application it is used in a strictly limited fashion since the weight requirement for the hydrogen containers is severe. It is generally used as a fuel only where its benefits can be maximized and in amounts that limit the unproductive baggage carried, namely, in later stages of multiple stage rockets. The use of complexing tertiary hydride equilibria as sinks for molecular hydrogen to reduce the weight of the containers merely substitutes the weight of the complexing metals for the steel and insulation for containers of liquid hydrogen.

The use of hydrogen as a fuel for land transportation is even less practical. Liquid–hydrogen fueled vehicles moving through urban centers present a high probability of violent and explosive collisions, risks that make this technology less desirable. The use of the complex hydride–hydrogen equilibrium sink offers little advantage over the rechargeable battery in weight to effective range. This would, at the present state of this technology, be an optimistic 50 miles between fillups at service stations furnishing hydrogen from liquid hydrogen storage facilities. Space limitations and energy-content-to-bulk ratios are not particularly advantageous even so. This already restricted range will be reduced dramatically by loss of time in conditions of regulated traffic. For portable fuels the hydrocarbons or their derivatives remain the most attractive option.

The state of our knowledge of the provenance and plenitude of fossil fuels is rudimentary. Indeed, most hydrocarbons may be "primordial" if Gold's hypothesis is correct.[46] At this point evidence seems to be mounting for its validity.[47] Analyses of gas samples from the East Pacific Rise, a spreading region of the ocean floor off the western continental shelf, demonstrated the presence of small quantities of methane and of large quantities of carbon dioxide.[48] Since the spreading ocean floor has no sedimentary deposits of biological origin to use in accounting for the methane, this observation has added credence to Gold's hypothesis. Similar evolution of methane is noted in other spreading areas of the ocean floor as well as in rift regions. Methane has been found in any ocean floor region of volcanic activity so far examined.[49] While this methane could be escaping from depths within the mantle, it also could arise in redox disproportionations to be discussed in this monograph. The disproportionation process is perhaps more useful here since it can accommodate logically the concurrence of oxidized (CO_2) and reduced (CH_4) compounds of carbon.

There is little logical basis to the assumption that the geological age of the rock in which a given deposit of petroleum (or coal, or oil shale, or tar sand) is found accurately assesses the time required for petroleum formation. Under appropriate conditions hydrocarbon formation from suitable carbon sources can be very rapid.[50] The chemistry discussed in later chapters suggests that such formation is virtually instantaneous. The chemistry discussed herein will also furnish a rational understanding of the conditions under which coal is formed in preference to hydrocarbons.

All that is required to support current and future industrial needs for energy, using this equilibrium chemistry as a source for liquid fuels, is a continuous supply of compounds with carbon in a reduced oxidation state. The photosynthetic carbon cycle guarantees such a supply so that, paraphrasing our native Americans, we need have no fears for energy depletion "as long as the waters run and the sun shines."

DIFFERENT APPROACH TO THE CHEMISTRY OF CARBON

The chemical study of compounds of carbon with hydrogen, oxygen, and nitrogen predates Wöhler's classic experiment[51] and gives the name *organic chemistry* to this discipline. Before that date "vital force" was deemed necessary to the formation of such compounds. With the Wöhler discovery, the vital-force argument collapsed and chemists rapidly began exploration of various reactions of these compounds. Over the hundred and fifty years that have elapsed the elegant and ordered structure of the field of organic chemistry has evolved.

With an abundance of these organic compounds immediately available from natural sources, the organic chemist has been interested primarily in the chemistry required to interconvert these many compounds. The study and identification of functional groups and the orderly organization of the chemistry characteristic of the different classes of compounds has resulted in an impressive edifice of scientific knowledge. A specialized language has developed from this science that greatly aids in creating an ordered discipline.

As techniques have improved, yet more complex examples of organic molecules have been characterized. Among the most complex are those molecules produced in the many biochemical processes and found to play vital roles in human well-being. Organic chemistry began to deal with these challenging complexities using the tools traditionally employed by their discipline. New physical techniques were applied to these problems as they became available. It is interesting that, in this effort, a modern counterpart of the vital-force concept was invoked. Special, almost mystical, roles frequently were proposed by scientists of the more biological approach. Not exactly as limiting as the historic vital-force arguments the modern equivalent can be more accurately called the "Ah! sweet mystery of life" mindset. Such mysteries usually were invoked when the role of enzymes were described. Understanding the chemical reactions involving enzymes is improving thanks to the discipline of organic chemistry.

With the overwhelming volume of science devoted to organic chemistry of all kinds, this monograph requires some justification. Without the abundant examples of compounds available to the organic chemist, the inorganic chemist approaches the problem of describing the chemistry of an element in a different way. A systematic examination of elementary chemical behavior is fundamental to this approach. The experimental effort of the world's many organic chemists has persuaded most inorganic chemists that the chemistry of carbon is perhaps adequately covered and their attentions are better directed to elements suffering from neglect. Were it not for the circumstance that most of our world's most useful compounds are those classed as organic, this decision would be supportable. There is, however, a paucity of chemistry devoted solely to the chemical behavior of elemental carbon. Buried among the huge literature of organic chemistry is a wealth of experimental observation and fact.

In the course of experimental work aimed at exploiting the photosynthetic carbon reduction in ways that could significantly reduce dependence upon petroleum, many interesting reactions have been observed. They are most accurately described as base-induced redox disproportionation reactions of carbon in its intermediate oxidation states. The products of these reactions can be altered by adjustment of conditions under which the equilibrium system is operated. The same chemistry is effected in the solid state employing processes similar to many currently used by chemical engineers in industry. It therefore enjoys a considerable advantage in producing classes of organic compounds that have wide utility without the necessity of greatly altering current technology.

The purpose of this monograph is to discuss this chemistry in a systematic and useful fashion against the rich historical background of organic chemistry. In this discussion the language will, of necessity, occasionally differ markedly from that used in the orderly field of modern organic chemistry.

REFERENCES

1. R. Carbo and A. Ginabreda, *J. Chem. Educ.*, **62**, 832 (1985).
2. L. W. Avery, in *Interstellar Molecules*, B. H. Andrews, Ed., Reidel, Dordrecht, 1980, pp. 47–58.
3. P. Freiberg, Å. Hjalmarson, W. M. Irvine, and M. Guelin, *Astrophys. Lett.*, **241**, L99–103 (1980).
4. W. M. Irvine, B. Höglund, P. Freiberg, J. Askne, and J. Elider, *Astrophys. Lett.*, **248**, L113–117 (1981).
5. A. Wooten, E. P. Bozyan, D. B. Garrett, R. B. Loren, and R. L. Snell, *Astrophys. J.*, **239**, 844–854 (1980).
6. H. E. Matthews, W. M. Irvine, P. Freiberg, R. D. Brown, and P. D. Godfrey, *Nature*, **310**, 125–126 (1984).
7. N. Broten et al., *Astrophys. Lett.*, **276**, L25–29 (1984).
8. A. Shimoyama, K. Harada, and K. Yanai, *Chem. Lett.*, 1183–1186 (1985).
9. J. Kissel and F. R. Krueger, *Nature*, **326**, 785–760 (1987).
10. R. Hyatsu, R. G. Scott, and M. H. Studier with R. S. Lewis and E. Anders, *Science*, **209**, 1515–1518 (1980).
11. A. G. Whittaker and E. J. Watts with R. S. Lewis and E. Anders, *Science*, **209**, 1512–1514 (1980).
12. Q. L. Zhang, S. C. O'Brien, J. R. Heath, Y. Liu, R. F. Curl, H. W. Kroto, and R. E. Smalley, *J. Phys. Chem.*, **90**, 525 (1986).
13. J. R. Heath, Q. L. Zhang, S. C. O'Brien, R. F. Curl, H. W. Kroto, and R. E. Smalley, *J. Am. Chem. Soc.*, **109**, 359–363 (1987).
14. S. R. Federman, A. C. Danks, and D. L. Lambert, *Astrophys. J.*, **287**, 219–227 (1984).
15. F. M. Devienne and M. Teisseire, *Astron. Astrophys.*, **147**, 54–60 (1985).
16. A. O'Keefe, S. McElvany, and J. R. McDonald, *Chem. Phys.*, **111**, 327–338 (1987).

17. Z. K. Alksne and Y. Y. Ikaunieks, in *Carbon Stars*, J. H. Baumert, Ed. (*Astronomy and Astrophysics Series*, Vol. 11, A. G. Pacholczyk, series Ed.) Pachart Publishing House, Tucson, Arizona, 1981, Chap. 2.

18. V. I. Kasotchin, A. M. Sladkov, Y. P. Kudravtsev, N. M. Popov and V. V. Korshak, *Dokl. Chem. (Engl. Transl.)*, **177**, 1031 (1967). See also: A. G. Whittaker and P. L. Kintner, *Science*, **165**, 589–591 (1969).

19. P. P. K. Smith and P. R. Buseck, *Science*, **216**, 984–986 (1982).

20. A. G. Whittaker, *Science*, **229**, 485–486 (1985); response by P. P. K. Smith and P. R. Buseck, *Science*, **229**, 486–487 (1985).

21. S. Misami and T. Kaneda, in *The Chemistry of the Carbon–Carbon Triple Bond*, S. Patai, Ed., Wiley, New York, 1978, Chap. 16.

22. R. Hoffmann, O. Eisenstein, and A. T. Balaban, *Proc. Natl. Acad. Sci. U. S. A.*, **77**, 5588 (1980). See also: R. Hoffmann, T. Hughbanks, M. Kertesz, and P. H. Bird, *J. Am. Chem. Soc.*, **105**, 4831 (1983); K. M. Merz, Jr., R. Hoffmann, and A. T. Balaban, *J. Am. Chem. Soc.*, **109**, 6742–6751 (1987).

23. J. P. Bradley, D. E. Brownlee, and P. Fraundorf, *Science*, **223**, 56–58 (1984).

24. L. S. Nelson, A. G. Whittaker, and B. Tooper, *High Temp. Sci.*, **4**, 445–477 (1972).

25. A. G. Whittaker and G. M. Wolten, *Science*, **178**, 54–56 (1972).

26. A. L. Lavoisier, *Dictionnaire de chymie*, Paris, Vol. 1, 504 (1792). See also: S. Tennant, *Phil. Trans.*, **87**, 123 (1797).

27. M. Tsuda, M. Nakajima, and S. Oikawa, *J. Am. Chem. Soc.*, **108**, 5780–5783 (1986).

28. H. P. Boehm, *Angew. Chem., Int. Ed. Engl.*, **6**, 535–538 (1966).

29. M. P. Crimmins and G. Urry, unpublished results.

30. M. Jeroviev and P. A. Latschinov, *J. Russ. Phys.-Chem. Soc.*, **106**, 1679 (1888); F. von Sandberger, *Neues Jahrb. Mineral., Geol., Palaentol.*, 171 (1889); W. Will and J. Pinnow, *Ber. Dtsch. Chem. Ges.*, **23**, 345 (1890); A. Daubrec, *Comptes rendus des Sci. de l'acad. dess.*, **110**, 18 (1890); E. Mallard, *ibid.*, **114**, 812 (1892); G. A. König and A. E. Foote, *Am. J. Sci.*, **42**(3), 413 (1892); C. Friedel, *Comptes rendus des Sci. de l'acad. dess.*, **115**, 1037 (1893); H. Moissan, *ibid.*, **116**, 218, 320, (1893); G. F. Kunz and O. W. Huntington, *Am. J. Sci.*, (3) **46**, 470 (1893).

31. A. P. C. Mann and D. A. Williams, *Nature*, **283**, 721–725 (1980).

32. M. E. Lipschutz and E. Anders, *Science*, **134**, 2095–2099 (1961).

33. M. E. Lipschutz, *Science*, **143**, 1431–1434 (1964).

34. R. S. Lewis, T. Ming, J. F. Wacker, E. Anders, and E. Street, *Nature*, **326**, 160–162 (1987).

35. F. J. M. Rietmeijer and D. R. Mackinnon, *Nature*, **326**, 162–165 (1987).

36. R. Roy, *Nature*, **325**, 17–18 (1987).

37. J. P. Wright and M. M. Grady, *Nature*, **326**, 739–740 (1987).

38. W. D. Huntsman, in *The Chemistry of the Carbon–Carbon Triple Bond*, S. Patai, Ed., Wiley, New York, 1978, Chap. 13.

39. F. Crick, *Life Itself*, Simon Schuster, New York, 1981.

40. C. Singer, E. J. Holmyard, A. R. Hall, and T. I. Williams, Eds., *A History of Technology*, Vol. 1, Oxford University Press, London, 1954.

41. M. Levey, *Chemistry and Chemical Technology in Ancient Mesopotamia*, Elsevier Publishing Company, New York, 1959.

42. *The Coming of the Age of Iron*, T. A. Wertime and J. D. Muhly, Eds., Yale University Press, New Haven, 1980.

43. H. Violette, *Ann. Chim. Phys.*, **23**, 475 (1848).

44. T. Lambert, *Bone Products and Manures*, London, 1901.

45. G. Urry and M. P. Santorsa, *U. S. Pat. App.* #166,869, filed Dec. 29, 1983; *European Pat. App.* #147,234, published Mar. 7, 1985.

46. T. Gold and S. Soter, *Sci. Am.*, **242**, 154–161 (1980). See also: D. Osborne, *Atlantic Monthly*, (Feb., 1986) pp. 39–54.

47. *New York Times*, **136**, 25, (Mar. 22, 1987).

48. J. E. Lupton and H. Craig, *Science*, **214**, 13–18 (1981).

49. R. F. Weiss, P. Lonsdale, J. E. Lupton, A. E. Bainbridge, and H. Craig, *Nature*, **267**, 600–603 (1977); J. E. Lupton, R. F. Weiss, and H. Craig, *ibid.*, **267**, 603–604 (1977); Y. Horibe, K. R. Kim, and H. Craig, *Nature*, **324**, 131–133 (1986).

50. J. D. Saxby and K. W. Riley, *Nature*, **308**, 177–179 (1984).

51. F. Wöhler, *Ann. Phys. Chem.*, **12**(2), 253–256 (1828).

Chapter II

Graphite

ANOMALIES IN THE BEHAVIOR OF GRAPHITE

In the course of studying equilibrium and nonequilibrium redox reactions of carbon in its intermediate oxidation states some interesting anomalies in the behavior of graphite were encountered that prompted an interest in a more detailed study of the chemistry of this form of elemental carbon.

For example, when graphite is mixed with solid KOH (15% water by weight) and the mixture melted *in vacuo*, hydrogen is irreversibly generated according to the stoichiometry of equation (1):

$$C + 2OH^- + H_2O = 2H_2 + CO_3^{2-} \qquad (1)$$

Since this reaction ensues at approximately $200°C$, graphite appears, under these nonequilibrium conditions, to be a highly reactive reducing agent. If, however, the same mixture is melted under equilibrium conditions (under autogenous pressure) in a closed tube and heated to even higher temperatures, little reaction occurs. Reaction of the elemental graphite can be promoted under these equilibrium conditions if the mixture is shaken vigor-

25

ously with a stirring or grinding ball free to move within the closed tube.   Under these conditions reactions like those in equations (2) and (3) prevail:

$$2C \;+\; OH^- \;+\; H_2O \;=\; CH_3COO^- \tag{2}$$

$$8C + 5OH^- + 3H_2O = CH_3CH_2CH_2CH_2CH_2COO^- + 2CO_3^{2-} \tag{3}$$

This behavior suggests that high reactivity and effectiveness as a reducing agent, similar to the nonequilibrium behavior *in vacuo*, can be promoted by grinding the graphite in the presence of the alkaline water solution of high ionic strength.

The chemistry of graphite represented in equations (2) and (3) can also be promoted in the equilibrium reaction with no grinding.   It is merely necessary to introduce, as an initiator for this reaction, a small amount of carbon monoxide gas at the outset of the reaction.

There is an overwhelming volume of literature dealing with the characterization of various forms of carbon.[1]   The major portion of that work applies physical methods in various attempts at surface character- ization.   Chemical studies with attempts to determine stoichiometry have mainly been directed to chemical characterizations of the more complex carbon blacks[2] and coals.[3]

Of all the forms of carbon, graphite is arguably that for which the most complete structural information is available. There is a virtual consensus that the basal plane structure is an extended two-dimensional hexagonal array of carbon atoms best described as a resonance stabilized polynuclear structure (the *ab* plane). There is less knowledge available for understanding the chemical nature of the prismatic faces (edges of the stacked *ab* planes). There is general agreement that these edges are oxygenated, but the precise nature of the oxygen substituents has not been elucidated.[4]

In the present work the methods of high-vacuum technique have been applied to the study of graphite in such a way that significant correlations can be demonstrated between chemical composition and chemical behavior. For example, if typical graphite is heated at temperatures just below $800^{\circ}C$, gas is evolved. If the heating is continued until no more gas is evolved and the pyrolysis, by the simplest and most valid of criteria, is complete, the gas can be measured accurately and its composition determined. For each mole of graphite, this mixture of gases consists of 0.30 mmol of carbon monoxide and 0.15 mmol of carbon dioxide. From these data it can be estimated that the composition of the graphite in question is approximately $C_{10,000}O_6$. This is an insignificant compositional difference rarely detectable by the statistically valid methods traditionally employed. Repetitions of this experiment give the same active oxygen content as well as the identical ratio of carbon oxides within ±1%.

The graphite produced by the pyrolysis displays chemical properties quite different from the original oxygen compound. It is, for example, pyrophoric and it reacts with water in a rational stoichiometry. It is possible to convince oneself that this small compositional change may account for the dramatic change in chemical behavior. This opinion is reinforced when this chemical behavior is related to the structure of the bulk of the graphite in a reasonable way.

It is possible to estimate the oxygen content in a different manner. Thus, if 1 mol of the same graphite is heated at temperatures just below $800°C$ in an atmosphere of carbon monoxide for several days, then cooled, and the gas removed and analyzed, the resulting mixture of gases is found to contain 0.57 mmol of carbon dioxide and 0.57 mmol less than the original carbon monoxide. The results of this chemical reaction confirm that the graphite in question has the composition, $C_{10,000}O_6$.

It is obvious that while the removal of oxygen from graphite by both methods accurately assesses the active oxygen content of the graphite, the evaporative method could possibly leave a different kind of graphite than the reductive method. Carbon is removed from the bulk structure in the evaporative case and not in the reductive case. Current knowledge of the bonding in the bulk of graphite should allow a correlation of such differences with the chemical behavior.

It is quite probable that ordinary graphite is unreactive for the same reason that ordinary aluminum is unreactive; because it is passivated by an oxide coat.[5] Any chemistry that removes this passivating layer enables the intrinsic reactivity of elemental carbon. Elemental carbon is "passive" towards potassium nitrate at ordinary temperatures. Yet when sulfur is added to a water slurry of these same reactants and carefully prilled and dried, it becomes the shock-sensitive gunpowder known for thousands of years. The reducing role of sulfur is indicated by the fact that another reducing agent, ammonium nitrate, can be substituted for the sulfur[6] of black powder to make smokeless powder. Ammonium salts other than the nitrate are also effective substitutes for sulfur.

Many similar illustrations of the high intrinsic reactivity of elemental carbon could be listed, including the reduction of alkali metal ions to alkali metals,[7,8] the reduction of molecular nitrogen in the classic cyanamide process,[9] and the many metallurgical applications that have been exploited since ancient times. These serve to support our present view.

GRAPHITE EDGE CHEMISTRY

The water-gas reaction[10] is most readily understood if the assumption is made that a high temperature is required in order to evaporate the oxide coating of the graphite structure and to expose the reactive edge of the elemental carbon. The reaction of water with freshly broken carbon has possibly been responsible for many of the coal-damp accidents in mines. We have studied all three of these methods of obtaining clean, or activated graphite: the evaporative or water-gas method, the reductive method, and the fracture method. It is fortuitous that the stoichiometry of these reactions is in the range where sufficiently precise determinations can be effected with high-vacuum techniques to shed light upon the chemical nature of graphite. The model that emerges from this chemistry is consistent with the oxygen passivation just described.

In this study we have laid the groundwork for a systematic derivative chemistry of graphite's edge. The initial assumption that ordinary graphite is stabilized by edge oxidation of the *ab* planes is well supported by the pyrolysis data just presented.

That no carbon compounds, other than oxides, are detected during this pyrolysis probably rules out edge stabilization by many other plausible organic functional groups. It does not rule out stabilization by pyridine or cinnoline edges, as we will discuss later in the preparation of nitrogen-passivated graphite. With the possible exception of the studies where typical graphite is oxidized by heating in air at $110°C$ we also are not able to comment rationally upon edge passivation involving carboxylic acid or acid anhydride groups or ethereal oxygen. The hydrogen-substituted graphite that we have prepared decomposes slowly at $800°C$ to generate hydrogen. This would appear to rule out any significant hydrogen substitution in our typical graphite. This and other edge derivatives will be described later in this chapter.

The temperature limitation forced upon us by the nature of quartz and its vaporization equilibrium,[11]

$$SiO_2 = SiO + \tfrac{1}{2}O_2 \tag{4}$$

has serendipitously demonstrated how the graphite produced by evaporation of its edge oxides is an avid oxygen scavenger. If the temperature of the graphite is allowed to rise to a point where any appreciable oxygen is produced by evaporation of silica, then no endpoint in the evolution of carbon monoxide can be reached. At the limiting temperature determined for our reaction the equilibrium pressure of O_2 can be estimated as approximately 10^{-4} torr.[12]

This has been somewhat limiting since it would be useful to proceed through the water-gas cycle in a stepwise fashion. Such a study requires a cycling through approximately $1000°C$. This cannot be done with the precision required for significant work since evaluating the oxygen arising from the silica equilibrium has proved too difficult to date.

In this discussion we will use drawings designed to illustrate the probable structural chemistry of the various reactions we have studied. By trial and error we have found that most of our chemistry and the structural changes accompanying this chemistry are suitably represented by an array of 78 carbon atoms (six staggered rows of 13). In order to compare the model reaction stoichiometries with the actual experimental values a ratio is determined between edge oxygen and bulk carbon atoms in the standard graphite. The best value for the molecular formula of an *ab* plane in the standard graphite is $C_{10,000,000}O_{6,000}$ assuming the accepted structure for graphite and our pyrolysis data. Thus the oxygen content in the model is approximately 500 times the oxygen content of the standard graphite. In the following discussions the equations have all been adjusted by this ratio to be comparable with the standard graphite. The ratios of various reactants in the sequential reactions are within the limits of experimental precision and provide a convincing match with the stoichiometries for the experimental examples described above. Of more consequence, the chemical behavior of the various product forms of graphite is consistent with current understanding of the structure.

Our best representation of a typical graphite is shown at the left in equation (5). There are simple structural consequences of the *ab* planar edge oxidation as illustrated. The main internal structure is locked in the quinoid form.

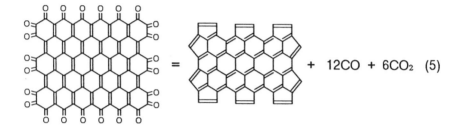

When this graphite is heated *in vacuo* at temperatures near 800°C, it loses carbon monoxide and carbon dioxide in the ratio of 2 to 1. The structure displays two different edges. The most likely source of the carbon monoxide is the 1,3-polyketone (horizontal) edges and that for the carbon dioxide is the 1,2- and 1,4-polyketone (vertical) edges.

It is a postulate of this model that no graphitization to regular six-membered rings occurs at these temperatures without the intervention of edge stabilizing reactants such as water, oxygen, or nitrogen. The fact that the loss of the oxides by evaporation proceeds to completion (using the simplest possible criterion that no more gas is produced) gives credence to a discussion of the reactive region of graphite. This region is most probably the edge of the lamellar planes.

This evaporative activation reaction occurs according to the stoich- iometry shown in equation (5). The stoichiometry is consistent with the formation of the activated graphite edged with fulvenoid rings and retaining an internal quinoid structure, illustrated as a product in equation (5). This requires biradical states along slightly puckered edges with 1,4- biradicals on the vertical and 1,2- or 1,5- biradicals on the horizontal. In the structural

equations, such as that above, there are possible resonance structures other than those presented for some cases. In the interest of
simplifying these complex equations, we show only one resonance
form, usually possessing edge carbon atoms that are trivalent, for
each case.

The recently characterized C_{60}, illustrated in Fig. 2.1, is
reported to be a spherical structure of joined five- and six-
membered rings that is remarkably unreactive.[13]   While the
characterization of this interesting compound has yet to be
confirmed, the melding of five- and six-membered rings seems to
have been accepted as a reasonable bonding arrangement.

The fulvenoid form for the active graphite proposed here is
not allowed the stabilization afforded C_{60} by the pairing of non-
bonding electrons distributed in a nonplanar structure.

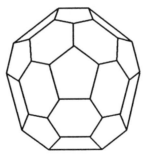

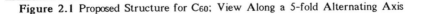

Figure 2.1 Proposed Structure for C60; View Along a 5-fold Alternating Axis

Such stabilization for the *ab* planes in the active graphite would come at a high cost in lattice energy. Edge puckering of the peripheral five-membered rings stabilizes this structure at less cost. Implicit in this puckering is an unpairing of electrons that helps rationalize the unusual reactivity of this form of graphite.

Fulvenoid graphite reacts with anaerobic water to form an interesting volatile product. One mole of carbon, activated by vacuum pyrolysis, reacts with an excess of water to give small amounts of ethanol and carbon monoxide. The resulting solid product is more stable thermally than is typical oxygenated graphite. The stoichiometry of this reaction is approximately that shown in equation (6):

$$+ \quad 22H_2O \quad = $$

$$+ \quad 6C_2H_5OH \quad + \quad 8CO \qquad (6)$$

In our model fulvenoid graphite there are vinylic biradicals along the horizontal edges. It is suggested that the ethanol arises from the hydrolysis of these vinylic biradicals leaving the benzenoid carbons along the horizontal edges oxygenated. The carbon monoxide arises from the vertical edges, leaving the remaining eight benzenoid carbons hydrogenated.

Heating water-passivated graphite at $800°C$ for several hours generates carbon monoxide and hydrogen in the ratio of 4CO to $1H_2$. This is possibly a low-temperature manifestation of the water-gas reaction. This form of passivated graphite is much more thermally stable than the typical graphite. The 1,2-diketones in typical graphite decompose readily to carbon dioxide. The hydrogen substitution in the water-passivated form precludes this mode of decomposition and helps to account for the enhanced thermal stability. However, this water-passivated graphite is not significantly different from standard graphite in its reaction with hot caustic *in vacuo*.

Fulvenoid graphite also reacts readily with carbon monoxide in an interesting reaction whereby the graphite is further reduced to an oxygen-free form that is more stable than the fulvenoid form. Reactions of various forms of graphite with carbon monoxide and carbon dioxide to other forms of graphite are discussed below.

The edge oxygen of typical graphite also can be chemically removed by a variety of reducing agents. Hydrogen has been used for this purpose, but it is likely that the graphite produced is hydrogenated. In solution, hydride complexes like formate ion effect this activation. Gaseous carbon monoxide at $600°C$ under equilibrium conditions can also produce an active form of graphite. If the graphite is equilibrated with pure carbon monoxide until no more carbon dioxide can be detected, the experimentally determined stoichiometry is in good agreement with that required by equation (7):

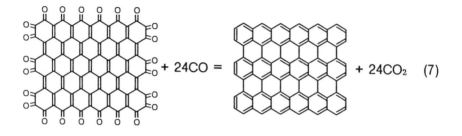

$$+ \ 24CO \ = \qquad\qquad + \ 24CO_2 \quad (7)$$

The reductive activation does not remove carbon as does the evaporative procedure. The remaining graphite is required to be in the benzenoid form. Here again, edge puckering will offer some stabilization. There is also implicit unpairing of electrons in such a puckered edge. The benzenoid product of equation (7) presents, on its horizontal edges, 1,3- biradicals and on the vertical edges 1,2- and 1,4- biradicals. The absence of five-membered rings on the edges would make this graphite less active than the fulvenoid form.

This reduction occurs in three distinct steps. The first stage of this reduction proceeds according to the stoichiometry of equation (8) and is essentially complete in 24 hrs. It is apparently completed before the later stages, discussed below.

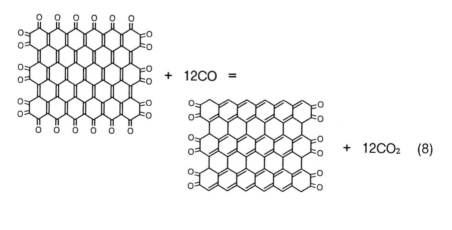

$$+ \ 12CO \ =$$

$$+ \ 12CO_2 \quad (8)$$

While the experimental data are not sufficiently detailed to allow estimation of rate constants for the different stages, the first and second stages of the reductive activation are complete in 48 hr and the second stage, as shown in equation (9), consumes no carbon monoxide.

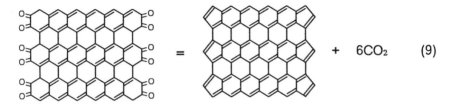

$$= \quad + \quad 6CO_2 \qquad (9)$$

After two weeks of equilibration at $600^{\circ}C$ the final conversion to benzenoid activated graphite is complete. An equation for this reaction that fits the precision of the experimental stoichiometry follows:

$$+ \quad 12CO \quad = \quad + \quad 6CO_2 \qquad (10)$$

Similar stoichiometry is observed in a reaction where fulvenoid graphite, prepared by the usual vacuum pyrolysis, is equilibrated with an excess of gaseous carbon monoxide for 18 hr at $800^{\circ}C$.

It is obvious from this reaction that the most stable form of oxygen free graphite at equilibrium with the oxides of carbon is the benzenoid form. This is in good agreement with the the fact that it is the least chemically reactive of any of the active graphites we have yet studied.

The charring of wood undoubtedly results in the formation of fulvenoid structures as the cellulose dehydrates at high temperature. The graphitization that occurs in charcoal burning also requires the conversion of these to the benzenoid form. Equilibria similar to the equilibria of equation (10) must also be pertinent to this process as well as to the graphitization accompanying the coking of coal.

The study of the reaction of benzenoid graphite with water has yet to be completed. When typical graphite is heated in the presence of a mixture of water vapor and carbon monoxide, a reaction occurs that may involve this form of active graphite. This reaction proceeds to completion in 18 hr at $400°C$ with the formation of carbon dioxide. This possibly involves concurrent reduction by the carbon monoxide and oxidation by the water vapor. The explosive oxidation of carbon monoxide, described by Davy,[14] is similarly catalyzed by water vapor. Moist air mixed with carbon monoxide displays wide explosive limits,[15] but dry oxygen will not react explosively with carbon monoxide.[16]

Equation (11) describes the stoichiometry for this water-catalyzed reaction of carbon monoxide with typical graphite.

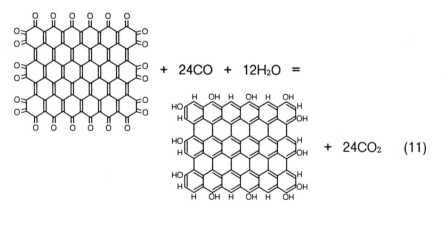

Typical graphite does not react readily with anaerobic water at 400°C. This method of water passivation promoted with CO is rapid compared to the preparation of benzenoid active graphite. The experimental stoichiometry, factored for the model, agrees well with that shown in the equation. The role of carbon monoxide in promoting this reaction is probably through reductive activation of some edge ketonic groups. The reoxidation of the active carbon atoms apparently proceeds autocatalytically in this case to produce the results as shown. The passivated graphite obtained in this reaction can be heated *in vacuo* to 800°C with little loss of carbon monoxide, unlike typical graphite.

The reaction product of equation (11) when slurried with water renders the water acidic with a pH of 5.0. While there is considerable resonance stabilization of this phenolic structure, it is susceptible to ready conversion to an unstable anionic form in the presence of bases. In an experiment that confirms this, passivated graphite displays no apparent differences from typical graphite in its reaction with hot concentrated caustic.

Such a conversion to a resonance-stabilized phenolic form may be the same process that results in the phenolic content of many coals. High phenolic content and its attendant hydration also helps to rationalize the surprisingly high water content of some western U. S. and Australian coals.

FIXATION OF NITROGEN USING GRAPHITE

Fulvenoid graphite reacts most readily with elemental nitrogen obeying the stoichiometry shown in equation (12). Benzenoid graphite reacts sluggishly in a more complex fashion. The graphite formed in the reaction with elemental nitrogen is very stable thermally and chemically.

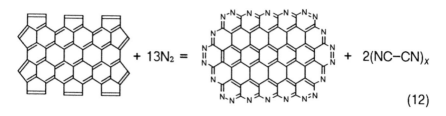

$$+ 13N_2 = \qquad\qquad + 2(NC-CN)_x \tag{12}$$

Fusing the nitrogen-passivated graphite product of equation (12) with a mixture of sodium and potassium hydroxides releases about 20% of its nitrogen content as ammonia. The cinnoline edge nitrogens are the most vulnerable to such treatment. Pyridine stabilization remains even after this reaction. Bone charcoal, used for thousands of years in applications where its enhanced chemical and thermal stability makes it superior to wood charcoal, may be similarly stabilized. Carbon fibers produced from the pyrolysis of polyacrylonitrile fibers are apparently stabilized by edge incorporation of nitrogen.[17]

This fixation of nitrogen is reminiscent of the preparation of sodium cyanide from the fusion of a mixture of sodium carbonate and carbon in air.[18] This has been studied extensively[19,20] and requires much higher temperatures than does the fixation by fulvenoid graphite. It is also strongly affected by silica impurities in the caustic.[20] The reaction of typical graphite with silica, which furnishes an understanding of this interference, is discussed later in this chapter.

An old wive's tale, learned by the time any rural subsistence farmer is old enough to pick berries, holds that a bumper blueberry crop is harvested on burned ground the year following a forest fire. The forests burned by these fires are usually evergreen forests. The soil under a growing evergreen forest is quite acid as a consequence of the rotting of fallen needles. Burning such a forest does produce enough alkali to remedy this. It would be possible to attribute the bumper crop of blueberries to this change were it not for the fact that the effect lasts for only one year and blueberries do not seemed to be harmed by acidity in the soil. Successful wild blueberries are to be found even in boggy areas where the soil is quite acid.

The effect is best described as a fertilization of some sort. It is possible that the conditions at the floor of such a forest fire closely approximate the conditions of the nitrogen graphite reaction, just described. The atmosphere should be mainly nitrogen and carbon monoxide. The temperatures should be high enough to char some wood to active graphite, which under these conditions might react with the prevalent nitrogen.

An experiment to prove this flight of fancy is not easy to conceive; however, the same results can be obtained by a youngster interested in increasing his or her disposable income. Judiciously applied barnyard manure, renewed each year, continues this blueberry dividend indefinitely.

The burning of stubble in an harvested field used to be a general farming practice to achieve similar results. The burning of sedge in boggy land was also widely employed. While the major benefit from this old established practice must be the adjustment of the soil pH, the nitrogen reaction with active graphite might well be a serendipitous result.

If this is indeed possible, the return to this practice should benefit farmers in a substantial fashion. Concerns about air pollution and acid rain have been largely responsible for forbidding such burning as well as the ancient ritual of autumn leaf burning. Millions of tons of alkali that might have been produced by this annual practice are not available. Is it possible that the deleterious effects of acid rain might have been mitigated were this restriction not imposed?

Certain generalizations can be drawn from the study of these various reactions of typical and activated graphites. Deoxygenated graphites are all pyrophoric to some degree. The most reactive inflame readily in air and react vigorously. Water vapor appears to assist the oxidation of active graphites in most cases. Whenever active graphites that are not in the benzenoid form are treated with anaerobic water, they form ethanol, carbon monoxide, and hydrogen as the main volatile reaction products. This does not appear to be the case when any oxygen, elemental or combined, is present, either in the graphite or in the gas phase in equilibrium with the graphite. In such case the active graphite is oxidized to an oxygenated form without any reduced products being formed.

REACTION OF ACTIVE GRAPHITE WITH ACETYLENE

Among the active graphites so far discussed fulvenoid graphite is the most useful in the preparation of new edge-derivatives. At room temperature the reaction with pure acetylene proceeds rapidly according to the equation:

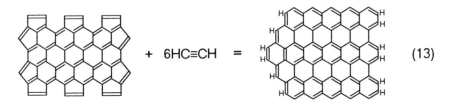

$$+ \; 6HC{\equiv}CH \; = \qquad\qquad\qquad\qquad (13)$$

The reaction just described sometimes requires a temperature as high as 200°C. This may be due to incomplete pyrolysis of the typical graphite leaving a small trace of oxygen in the active graphite.

The reactions of acetylene with the same form of active graphite, which occur at more highly elevated temperatures, are informative. Under such conditions free-radical pathways become important; the graphite consumes more acetylene and produces higher alkynes such as diacetylene:

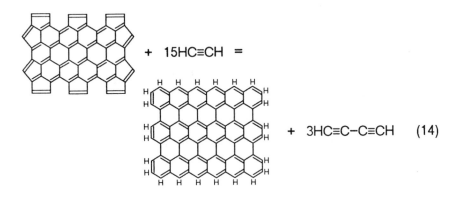

$$+ \; 15HC{\equiv}CH \; =$$

$$+ \; 3HC{\equiv}C{-}C{\equiv}CH \qquad (14)$$

At 400°C the stoichiometry shown in equation (14) dominates, with the graphite product now fully hydrogenated and fully stabilized. The graphite product, while not as stable as nitrogen-passivated graphite, is considerably less reactive and more stable thermally than typical graphites. This is an expected consequence of the resonance stabilization afforded by a fully benzenoid structure.

At temperatures above 400°C more acetylene is consumed and hydrogen is evolved essentially in equimolar quantities. In this stage of the reaction carbon is being added to the graphite lattice without any alteration of its edge hydrogenation. This is an orderly counterpart of the thermal decomposition of acetylene[21] observed in an oxyacetylene torch supplied with insufficient oxygen. At these elevated temperatures small amounts of carbon monoxide are also observed.

The carbon monoxide probably results from a side reaction of the diacetylene product of equation (14) with the quartz reaction vessel, since the conditions of the experiment allow this product to accumulate as the acetylene is consumed. These traces of carbon monoxide are accompanied by other alkynes, such as propyne and 2-butyne, which may also result from this side reaction. When the reactant gas is pure acetylene no carbon monoxide is found among the products.

Other acetylenes such as 2-butyne and hexafluoro-2-butyne react with fulvenoid graphite at room temperature with a stoichiometry similar to that expressed in equation (13). The reaction at 200°C with both of these substituted alkynes is more complex. It appears that approximately stoichiometric amounts of diacetylenes are formed with the implication that the edge substitution has been completed as in equation (14). Some alkanes and alkenes are also found among the volatile products at this temperature. The current rationale for this suggests that methyl and trifluoromethyl edge substituted graphite both pyrolyze at this temperature with the generation of methylene or difluoromethylene. In each case the volatile products other than the diacetylenes can be described as secondary products of these methylenes. Methane, ethane, and ethylene are generated as by-products of the 2-butyne reaction with fulvenoid graphite. In the case of the hexafluoro-2-butyne reaction perfluoromethane and tetrafluoroethylene are noted among the products along with a complex mixture of other volatile fluorocarbons.

If commercially obtained acetylene is used in these reactions, an interesting difference is noted. A great deal more acetylene is consumed at room temperature. The difference in this reaction is possibly a consequence of an unmeasurable impurity of acetone coming from the acetylene storage tank. Pure acetylene, purposefully contaminated with a trace of acetone, also reacts in this fashion. There is further work to be done on the unusual catalytic activity of acetone on this as well as other reactions of active graphite.

The stoichiometry of the reaction with tank acetylene requires there to be hydrogen other than aromatic hydrogens in this structure. There is no graphitic structure that can be drawn without some form of alkyl substitution. The model equation is shown with the methyl substituents symmetrically arranged. There is no experimental basis for this, and it is done solely for ease of representation. It is likely that the methyl substitution is random or alternate.

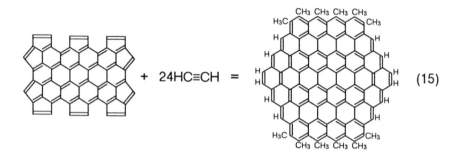

$$+ \ 24HC\equiv CH \ = \tag{15}$$

An alternate rationale for the stoichiometry observed for this reaction would involve a product graphite with a dramatically different edge-to-bulk ratio than that for typical graphite. To achieve the hydrogen content of the product of equation (15) by simple edge hydrogenation of a benzenoid graphite would require at least a tenfold reduction of average particle size for the resulting graphite. No such particle size reduction for the product graphite is evident. Further, pyrolysis of this product graphite gives rise to gaseous hydrocarbons that are most easily understood as arising from the thermal decomposition of edge methyl substituents.

Since this reaction occurs so smoothly and rapidly at ambient temperatures, it is unlikely to involve any free-radical processes. In order for methyl or other alkyl substituents to be formed, the protons in the course of this reaction must be free to assume minimum potential positions. This strongly suggests some form of tunneling process in intermediates for this reaction.

This unusual result prompted a reexamination of the reaction of pure acetylene with fulvenoid graphite. After an initial rapid reaction with the stoichiometry of equation (13) that is essentially complete in 2 hr, a very slow uptake of acetylene continues. After four weeks at room temperature the stoichiometry approaches that of the tank acetylene reaction. The rapid reaction in the presence of acetone is clearly catalytic.

REACTION OF GRAPHITE WITH QUARTZ

The quartz vessels used for the high temperature reactions in the vacuum apparatus frequently have some of the graphite product sintered to the inside quartz surface. This creates a practical problem in preparing the vessel for reuse. A lazy person's way of cleaning this quartzware is to subject it to an annealing cycle, where it is heated in air for several hours at temperatures near $600°C$. In most cases this suffices to burn the graphite sinter away, leaving a clean vessel. Both the nitrogen-passivated and fluorocarbon graphites are refractory in this procedure and require extraordinary oxidation times to complete the air-oxidation. This also is a measure of their chemical and thermal stability. In some instances it is necessary to wash the quartzware with dilute aqueous hydrofluoric acid in order fully to remove the last traces of carbon.

Refractory graphites strongly bonded to the quartz vessel are formed when pyrolyses of typical graphite are carried out at temperatures in excess of $800°C$. In these cases it also is not possible to reach an endpoint in carbon monoxide evolution. Equation (4) was offered as possibly accounting for this continued oxidation of the graphite.[11]

The formation of the chemically stable graphites in these cases suggested an experiment where an intimate mixture of high surface silica powder was mixed with finely ground graphite. This mixture was heated for several hours at $900°C$ until no further evolution of gas could be detected. While the silica was present in excess for this reaction, the stoichiometry of carbon monoxide generated to graphite used in the experiment agrees with equation (16):

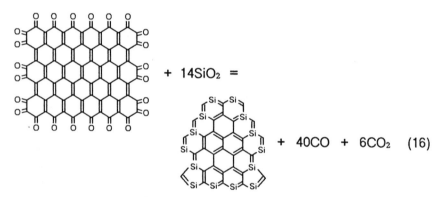

$$+ \ 14SiO_2 \ = \qquad\qquad + \ 40CO \ + \ 6CO_2 \quad (16)$$

In this reaction all of the carbon dioxide and 30% of the carbon monoxide is evolved by pyrolysis before the temperature exceeds $800°C$. This indicates that the vacuum pyrolysis of equation (5) is complete before any reaction with silica commences. It is best described as a reaction of fulvenoid graphite with silica as illustrated in equation (17):

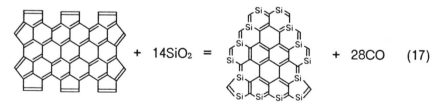

$$+ \ 14SiO_2 \ = \qquad\qquad + \ 28CO \quad (17)$$

The structure as presented for the silicon carbide-edged product includes two fulvenoid rings. No implications of unusual reactivity are intended. A fully benzenoid structure is the likely result of this reaction, but to illustrate such a reaction would require structures covering four times the areas of those used here. The fulvenoid rings are merely an artifact of the size of the structural model used for convenience in this case.

The graphite product of this reaction is refractory and quite inert chemically. It is possibly responsible for the interference by silica in the high-temperature Bucher process for the synthesis of sodium cyanide.[19] Prolonged oxidation in an annealing oven merely creates a white frosting on the surface of the solid. This may be silica but it is more likely to be the result of edge oxidation to form silicone linkages that are interlayer and/or intercrystalline. Such linkages are possibly responsible for the tenacious bonding to the silica vessel. Washing the coated quartz vessel with dilute hydrofluoric acid frees the graphite from the surface, but apparently does not dissolve the graphite particles so freed. The chemical characterization of this material is also difficult as a consequence of its chemical stability. To date, no analytical method has been useful in this characterization.

OXIDATION OF OXYGEN-PASSIVATED GRAPHITE

The further air oxidation of typical oxygen passivated graphite also is rational according to our model, apparently producing yet another different edge substituted graphite. At the modest temperature of $110^{\circ}C$ in an air oven the oxidation is fairly rapid. The results are readily interpreted as the conversion of the ketonic edge substituents to carbon dioxide and phthalic anhydride moieties. A water slurry of this graphite is acidic.

The vacuum pyrolysis of this graphite produces greater amounts of carbon dioxide, consistent with an increased active oxygen content. The relative amounts of carbon monoxide and carbon dioxide change as a function of the length of time a sample is heated in the air oven at $110^{\circ}C$. After several weeks of air oxidation at this temperature, vacuum pyrolysis yields only carbon dioxide along with a trace of water. At this point the edge substituents must be mainly carboxylic acid anhydride groups with occasional carboxylic acid moieties. The loss of weight of the graphite corresponds with that required for the stoichiometry of equation (18):

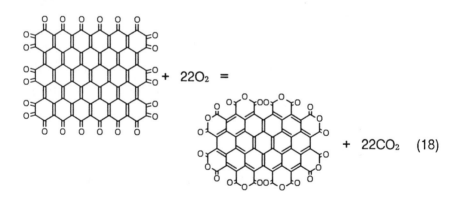

An oxidation by moist air in a closed system also generates carbon dioxide in an amount that agrees with the equation as written.

The structure for the product graphite of this reaction offers a reasonable driving force for this oxidation. The quinoid form of the fresh graphite is modified so that the edge of each *ab* plane is benzenoid. The bulk internal structure of the planes, however, can be seen to remain in the quinoid form.

This oxidation apparently occurs slowly even at ambient temperatures. A sample of the 99.9% graphite has been observed to degrade over a period of 14 months as indicated by a gradual decrease in the ratio of CO to CO_2 produced upon vacuum pyrolysis along with an attendant increase in apparent oxygen content. Graphite, aged over a long period after the container is opened initially, also gives an acidic reaction with water. A freshly opened sample always gives a carbon-monoxide-to-carbon-dioxide ratio of 2 to 1 upon pyrolysis.

The active fulvenoid form of graphite clearly is capable of many other reactions that could result in other interesting edge derivatives. While the stoichiometric chemistry discussed so far in this chapter produces quantities of volatile products that are uninteresting as a synthetic method, it results in the formation of novel graphites that display different and interesting properties. There are obvious applications in the areas of ceramic, lubricant, and carbon fiber technologies.

TRIBOLYTIC GRAPHITE

There is another type of highly reactive graphite, variously known as pyrophoric or tribolytic graphite, that is generally produced by the grinding of typical graphite. When ground extensively in air, naturally occurring graphite is converted to a form in which there has been considerable displacement of the *ab* planes. These fragmented planes apparently are edge-bonded one to another to form an extended open structure. It is likely that the edge-bonding bridges are ethereal oxygens or carboxylic anhydride moieties. Steric requirements would leave some edge positions hindered and unsubstituted, accounting for the enhanced reactivity of this form of active graphite.

Considering the most probable consequences of this milling or grinding of typical fresh graphite, it is likely that most of the particle size reduction will be mainly a consequence of lamellar cleavage. Any activation of the graphite structure, according to our model, will require fracture of the *ab* planes. Such fracture could occur in the shear of one plane slipping past another and would largely be confined to the edges of the lamellar planes. The internal quinoid structure of typical graphite naturally predisposes the three modes illustrated in Fig. 2.2. This is a direct consequence of bond energy differences.

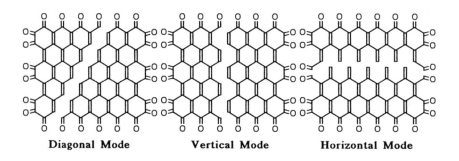

Figure 2.2 Modes of Tribolytic Fracture of Graphite's *ab* Plane

The horizontal mode is a higher energy mode, requiring more bond breaking and yielding fragments considerably higher in energy than the other two modes. The remaining modes are equally probable energetically. The edges produced by such fracture are the 1,3-polyketone and the 1,2-;1,4-polyketone edges that the stoichiometry of pyrolytic decomposition requires.

Since they are equally probable, the observation of the 2-to-1 ratio of carbon monoxide to carbon dioxide in the pyrolytic decomposition is a natural consequence of the grinding process. The structural models used in the equations of this chapter are rectangular. To be more properly descriptive of ground graphite they should be represented as parallelograms. Such a representation would require much more page space without any benefit in illustrating the stoichiometry of the various reactions.

Quantitative studies of reactions of tribolyzed graphite are difficult and require the construction of a closed system apparatus in which vapors or liquids can be brought into contact through a range of temperatures with graphite in the process of being ground. There are no data currently available from such experiments. It is to be hoped that this deficiency will be overcome in the not-too-distant future.

Qualitative experiments have produced results that confirm an unusually high reactivity for this form of activated graphite. When a sample of typical graphite is ball-milled with agate pebbles for 24 hr at ambient temperature, the resulting graphite demonstrates a composition markedly different from typical graphite. Vacuum pyrolysis of this pulverized graphite at $800°C$ produces a complex mixture of alkanes, alkenes, and alkynes, with methane, ethane, and acetylene forming the major portion of the volatile product. Carbon monoxide is also a product of this pyrolysis as it is in the pyrolysis of typical graphite.

This procedure is carried out essentially in the open atmosphere with fortuitous water vapor present. It is likely that the tribolyzed graphite is reacting with this water vapor to produce edge-substituents that pyrolyze to the hydrocarbon products observed. This is the most likely source of hydrogen in the reaction scheme. The quantities of hydrocarbons produced upon vacuum pyrolysis appears to correlate with the length of time the graphite is milled, with longer milling times producing greater quantities of hydrocarbons. A reaction similar to this could occur in working coal mines as the source of coal damp.

This tribolytic activation of graphite has helped to account for the behavior on one occasion of a new sample of freshly opened graphite powder. The container was opened and the reaction vessel loaded rapidly. Upon pyrolysis this sample demonstrated an anomalously low oxygen content and produced small amounts of hydrogen during the pyrolysis. A repetition of the experiment loading the reaction vessel from the same sample of graphite, opened a day or two previously, gave results that did not display these anomalies. It is likely that the supplier packaged this sample of powdered graphite before it had equilibrated with the ambient air.

A simple procedure promises to yield quantitative information for these tribochemical reactions of typical graphite with liquids. A completely degassed liquid is placed with fresh powdered graphite in an evacuated glass vessel along with three or four grinding balls of Hastelloy C^{TM}. The sealed vessel containing the reaction mixture is placed in an ultrasonic water bath and agitated for several days. At the end of this treatment the reaction vessel is attached to the vacuum apparatus, and opened, and the material volatile at temperatures up to 200°C is removed and analyzed. In one case, where the reactant liquid was water and the agitation period one week, carbon dioxide and acetic acid were found among these volatile products. The amounts formed were so small as to indicate that this procedure will require very long ultrasonic treatments in order to get product mixtures in sufficient quantities to effect good quantitative separations and characterizations. In the absence of the grinding balls no products were detected.

When the reactant liquid was methanol, acetaldehyde was found as a product instead of acetic acid. Vacuum pyrolysis of the graphite at 800°C, after all of the reactant methanol had been removed previously at 200°C, produced a mixture of carbon monoxide, hydrogen, and methane.

As more effective apparatus and procedures are developed where extensive grinding can be accomplished in the presence of liquids and gases of choice over a wide range of temperature, tribochemical processes could yield useful amounts of products. The energy consumed in such processes is low, and such reactions promise to furnish highly practical results.

REFERENCES

1. J. M. Thomas, *Carbon*, **8**, 413–421 (1970); see also: *Chemistry and Physics of Carbon*, vols 1–19, P. L. Walker and P. A. Thrower, Eds., Marcel Dekker, New York, 1966–1984.

2. D. S. Villars, *J. Am. Chem. Soc.*, **69**, 214–217 (1948); *ibid.*, **70**, 3655–3659 (1970); A. Swiatowski, *Z. Phys. Chem., (Leipzig)*, **265**, 1026–1033 (1984).

3. T. Green, J. Kovac, D. Brenner, and J. W. Larsen, in *Coal Structures*, R. A. Meyer, Ed., Academic Press, New York, 1982; M. Shibaoka and N. J. Russell, *Fuel*, **62**, 607–610 (1982); C. J. Chu, S. A. Cannon, R. H. Hauge, and J. L. Margrave, *Fuel*, **65**, 1740–1749 (1986).

4. H. P. Boehm, *Angew. Chem., Int. Ed. Engl.*, **6**, 535–538 (1966); J. M. Thomas, *Carbon*, **7**, 359–364 (1969).

5. J. M. Thomas and E. E. G. Hughes, *Carbon*, **1**, 209–214 (1964).

6. T. L. Davis, *Chemistry of Powder and Explosives*, Vols. 1 and 2 (combined), Wiley & Sons, New York, 1943.

7. F. R. Curadau, *Ann. Chim. (Paris)*, **66**, 97–103 (1808).

8. J. L. Gay-Lussac and J. L. Thenard, *Recherches Physico-Chimiques*, Paris, **1**, 101 (1811).

9. A. Frank and N. Caro, Brit. Pat. #15066 (1895).

10. F. Clément and J. B. Désormes, *Ann. Chim. Phys.*, **38**, 285 (1801). See also: J. J. Morgan, in *Chemistry of Coal Utilization*, Vol. 2 H. H. Lowry, Ed., Wiley, New York, 1945, pp. 1673–1749.

11. L. Brewer and D. Mastick, *J. Chem. Phys.*, **19**, 834–843 (1951).

12. L. Brewer and R. K. Edwards, *J. Chem. Phys.*, **58**, 351–358 (1954).

13. Q. L. Zhang, S. C. O'Brien, J. R. Heath, Y. Liu, R. F. Curl, H. W. Kroto, and R. E. Smalley, *J. Phys. Chem.*, **90**, 525–528 (1986).

14. H. Davy, *Phil. Trans. Roy. Soc. London*, **104**, 557 (1814).

15. E. Berl and H. Fischer, *Z. Elektrochem.*, **30**, 29–36 (1924).

16. H. B. Dixon, *Ber. Dtsch. Chem. Ges.*, **38**, 2419–2446 (1905).

17. Commission on Polymer Characterization and Properties, *Pure Appl. Chem.*, **58**, 455–468 (1986).

18. L. Thompson, *Mech. Mag.*, **No. 822**, 92 (1839).

19. J. E. Bucher, *Ind. Eng. Chem.*, **9**, 233–253 (1917).

20. E. W. Guernsey, J. Y. Yee, J. M. Braham, and M. S. Sherman, *Ind. Eng. Chem.*, **18**, 243–248 (1926).

21. R. S. Slysh and C. R. Kinney, *J. Chem. Phys.*, **65**, 1044–1045 (1961).

Chapter III

Carbohydrates

INTRODUCTION

Photosynthetically active organisms and plants provide a virtually
constant and renewable source of compounds with carbon in its
elemental oxidation state. This fortuitous state of affairs is the
result of the photosynthetic cycle which converts carbon dioxide
to glucose and oxygen.[1] Other biochemical processes convert
glucose to the various and diverse compounds necessary to the
functioning of the plant or other organism. The natural products
produced by plants include many compounds with useful and
marvelous properties. Recently, only those capable of medicinal
applications have been studied with any intensity. The rapid
development of the sub-discipline of *natural product synthesis* is a
measure of the value accorded these medicinal plants.

Carbohydrates in which no change in the oxidation state of
the carbon occurs range in complexity from simple sugars and
polysaccharides, major components in the cell walls of simple
algae, to more complex polysaccharides such as starches, and
cellulose. Starches are necessary for plant respiration and are
stored in seeds as nutrition for plant embryos in seeds, and
cellulose is the structural material used for mechanical support in
larger plants and trees.

The carbohydrates constitute a class of compounds that enjoy a wide range of useful chemical and physical properties, most of which arise from a lower degree of hydration and a consequent higher degree of polymerization.   Uses as food involve more subtle differences.   The number of carbons in a sugar and differences in stereochemical configuration as well as the size of the lactone ring produce differences in flavor, nutritive value, and physical behavior.   The reactions involved in conversions of primary glucose are simple but specific dehydrations.   These dehydrations occurring within biological systems must be effected in such a way that stereochemical requirements are also satisfied. Much of the complexity of plant biochemistry arises from these requirements.   Racemic hydrations and dehydrations are not nearly so complex. Complete dehydration of any carbohydrate to elemental carbon is trivially simple.   Hydration of elemental carbon to various carbohydrates can be similarly simple.   Emil Fischer used simple chemical reactions to gain much of our present detailed knowledge of saccharides and built an elegant intellectual edifice that appropriately earned a Nobel Prize.[2]

OBSERVATION OF CARBOHYDRATE BEHAVIOR

Plant carbohydrates —sugars, starches, and celluloses— have been studied formally and experientially for the longest period. Bread making, candy making, wine making, and other related domestic practices produced chemical knowledge before the establishment of a formal chemical science.   Most humans early in life assimilate a considerable store of chemical knowledge about this class of compounds since it constitutes one of the main food groups for animals.

Fermentation of fruit juices to more interesting beverages with different flavors, such as the mild carbonation and hardening of apple cider, is observed naturally in the early experiences. Distillation of fermented water solutions to enhance the concentration of the alcohol was surely one of the earliest examples of chemical separation by physical means. To some, the odors associated with the fermentation chemistry of bread making and baking are among the most poignant olfactory memories. It is the rare individual who proceeds to maturity without observing the charring or dehydration of sugars in oven spills or enjoying the pleasure of converting too sweet glucose to pleasant tasting caramel candy. Candy making offers the first opportunity to observe the effects of acids, such as vinegar or cream of tartar, as well as bases, usually baking soda, in altering the properties of the final product.

Anyone tending herbivorous animals learns of the chemical stability of cellulose. After passage through the digestive system of these animals the cellulose remains largely unaffected. Cleaning the Augean stables was firmly established as a Herculean task in mythology and modern automation does little to make this task more pleasant. Even the ungulates with their ruminant digestive system, assisted by gastric microorganisms, are unable to digest all cellulose. This chemical stability is also responsible for one of the current uses of cellulose. Very large quantities of refined wood cellulose are added to foods. The purpose is to add nonnutritious bulk to many diet foods and cereals.

The cellulose hulls of some seeds are so chemically inert that they require unusual treatments before they are sufficiently permeable to water to allow germination. The seed of the allspice or pimento tree will not germinate unless it is treated with alkali. In nature this treatment is effected in the highly alkaline digestive tract of a bird. Seed hulls, in the main, are comprised of celluloses that are polymers of pentoses, the tetrahydroxypentaldehydes. The well-advertized bran is such a cellulose. Much of the chemistry of the plant carbohydrates has been ascertained from similar homely observation.

Among the earliest chemical fascinations is that of burning cellulose, whether in wood or paper. Man's early dependence upon the benisons of fire, combined with the all too frequent experience of its destructive potential, undoubtedly produced this preoccupation. Intrinsic to this preoccupation is a primal knowledge of the nature of charring and combustion. Ritual uses of fire, including the Guy Fawkes bonfires in Britain and the offertory candles of many churches manifest the compelling nature of this fascination.

It is possible for anyone to observe the chemical consequences of adding water vapor to the burning of cellulose. The addition of wet wood to a well-established wood fire, whether in a home fireplace or in a campfire, produces results related to this phenomenon. As the heat of the fire drives water vapor from the wet wood, flares of flame appear. These flares are the direct consequence of the formation of volatile products in the water-assisted pyrolysis of cellulose. These burst into flame as soon as they encounter sufficient oxygen to support incandescent combustion.

CARBOHYDRATES— DEFINITION AND CLASSIFICATION

It has proved useful in the discipline of organic chemistry to order carbohydrates into classes that reflect different conditions of polymerization or dehydration. There is virtual agreement in advanced textbooks that a useful definition of carbohydrates is: ". . . polyhydroxy aldehydes or polyhydroxy ketones *or substances that yield these compounds upon hydrolysis.*" The simple sugars, or monosaccharides are either aldoses like glucose or ketoses like fructose, the sweet component of most fruit juices. Disaccharides are typified by sucrose, a mixed anhydride of glucose and fructose. This carbohydrate is the purest chemical compound produced in such great quantities. Typical annual production is many billions of pounds. Polysaccharides like starch are important foods and because of their inexpensive production in farming are useful industrial chemicals. Cellulose, the most highly polymerized carbohydrate, finds many engineering and structural applications in the form of wood and its derivatives.

The term *carbohydrate* is linguistically devised to group together those compounds that can be viewed as hydrates of elemental carbon. While general usage would not group acetic acid in this class of compounds, the classification will be used in this monograph in its literal sense. Thus, acetic acid, methyl formate, glycolic aldehyde, $C_2(H_2O)_2$, and pyruvic aldehyde, $C_3(H_2O)_2$, are carbohydrates under our definition. A nomenclature that is descriptive of the oxidation state of the carbon content of a compound will prove more useful in the discussions to follow.

In Table 1.2 and its references we find several interstellar molecular species that are carbohydrates or would readily hydrate to form derivatives of carbohydrates. Ketene and higher members of its homologous series, $C_{2n}H_2O$ constitute carbohydrates of minimal hydration. The series C_nH_2O with the lowest homolog formaldehyde constitutes the series of maximal hydration. Between these two extremes lie many reactive and synthetically fertile series of compounds.

DEGRADATION OF CELLULOSE

Modern scientific interest in the burning phenomenon continues and a recent report on the products of flash pyrolysis of microcrystalline cellulose identifies the wealth of compounds burning in such flames.[3] The many different compounds identified in this study are listed in Table 3.1.

The low yields of reduced product in this table should come as no surprise. The hydrogen content of cellulose is only 1.2% by weight.

Table 3.1 Compounds from Flash Pyrolysis of Cellulose at 310°-770°C

Compound	Yield as Percent of Cellulose Consumed
Carbon monoxide	60
Carbon dioxide	7.0
Glucosan tars	9.5
Water	9.5
Acetaldehyde	1.1
Methane	1.0
Ethane-ethylene	~1
Propane-propene	~1
Butane-butenes	~1
Furfural	0.5
Acetone	<1
Methanol	<1
Acrolein	<1
Crotonaldehyde	<1
Furan	<1
Benzene	<1
Butyric acid	<1
2-Methylfuran	<1
Toluene	<1

Volatile oxides listed in the table account for 85% of all the available oxygen and 64% of the available carbon. The estimated amounts of reduced product reported would, in fact, appear to be too high.

The behavior of carbohydrates in alkaline aqueous solution is of primary interest in this monograph. In dilute solutions aldoses are in equilibrium with ketoses. In the de Bruyn–van Ekenstein transformation,[4] calcium hydroxide at room temperature apparently converts aldoses to ketoses without achieving equilibrium. Other reactions are observed in potassium hydroxide solutions in the range of 10^{-3} to 10^{-2} molar hydroxide ion. Aldol condensations and water oxidation to saccharinic acids and oligomers of these acids dominate at these OH^- activities. At higher activities in moderate concentrations alkaline degradation accompanies water oxidation with lactic acid being a principal product.[5]

In more concentrated caustic extensive carbohydrate degradation occurs. An interesting hydroxydiketone, 3-hydroxy-2-keto-propionaldehyde (reductone), is produced in some such degradations. Although it is more highly oxidized than a carbohydrate, it is apparently a powerful reducing agent. Formaldehyde is also reported among the products of alkaline degradation. This is puzzling since formaldehyde would certainly be expected to undergo condensations or oxidation to formic acid or bicarbonate under such conditions. John U. Nef spent a major portion of his active career examining the behavior of reducing sugars in the presence of strong bases.[6] A useful review by W. L. Evans[7] summarizes the reactions of reducing sugars in alkaline solutions.

Table 3.2 Low Molecular Weight Compounds from Degradation of Cellulose at pH 12.50 and 300°C

Compound	Yield as Percent of Total Oil Product
2,4-Dimethylfuran	0.50
2-Methylcyclopentanone	0.30
Phenol	0.20
3-Heptene	0.20
Cyclohexanone	0.10
2,5-Dimethyltetrahydrofuran	0.10
m- or p-Cresol	0.10
Cyclopentanone	0.10
o-Cresol	0.09
2-Ethylcyclopentanone	0.09
2,5-Dimethyl-2,4-hexadiene	0.09
4-Methyl-3-penten-2-one	0.06
2,5-Dimethyl-2-cyclopentenone	0.06
p-Cresol	0.03

Recently, low molecular weight compounds from the oil produced by cellulose degradation with an aqueous sodium carbonate solution (pH 12.50) in an autoclave at 300°C have been analyzed and identified.[8] Those compounds which have been reasonably well identified are listed in order of decreasing yields in Table 3.2.

The autoclave conditions under which the cellulose was converted to an oil yielding the components listed in Table 3.2 did not allow retention of any gases, such as carbon monoxide, carbon dioxide, hydrogen, methane, ethane, ethylene, or acetylene. Compounds with low solubility in water possessing vapor pressures exceeding the control pressure of the autoclave were also lost. This experiment differs in one important respect from the flash pyrolysis summarized in Table 3.1. There is essentially unlimited water available to the conversion reactions summarized in Table 3.1.

The advantage of increased water activity in converting cellulose to liquids similar to petroleum is exploited in high pressure steam conversions.[9]

The high incidence of unsaturated compounds listed in Table 3.2 gives evidence of considerable dehydration under the conditions of this experiment. The presence of a char, presumably elemental carbon, upon completion of the reaction is further evidence of this dehydration, but it also indicates a water activity too low to support higher conversions. A number of other products were tentatively identified in this study. The list as it stands gives solid evidence for redox—disproportionation reactions producing compounds similar to those produced in living plants.

It is not possible from the data given in this work to determine the fraction of the carbon present in the reactant cellulose that is converted to liquid products. An accurate estimate would not be possible without data for the gaseous carbon-containing products not collected in this experiment.

INDUSTRIAL USE OF CELLULOSE

Cellulose has been an important industrial chemical since the middle of the nineteenth century. In addition to the large volume of use as a food additive its conversion to clothing fibers and wrapping films are important industrial processes.

Fully nitrated cellulose was first prepared [10] in 1846, and 20 years later it was being spun into fibers from highly viscous solutions in alcohol—ether mixtures. These solutions, known as collodion, were also used as a fast plastic cement. An incompletely nitrated cellulose, pyroxylin, found the widest use in the manufacture of fibers and plastics. After the fibers were spun, the nitrate was removed by a solution of sodium bisulfide. Fully nitrated cellulose is the major constituent of guncotton. A gel of guncotton dissolved in nitroglycerine was first prepared by Alfred Nobel. This blasting gelatin is safer than nitroglycerine and is similar to dynamite in shock stability.

The explosive properties of nitrated cellulose and the expense of using a natural fiber as a starting material for a synthetic fiber prompted development of a process based upon the cheaper wood cellulose. In the viscose process, developed at the end of the 19th century, wood cellulose is first made alkaline with dilute sodium hydroxide. The sodium salt is then treated with carbon disulfide to form the dithiocarbonate or xanthate, named for the intense yellow color of these salts. The xanthates form a viscous colloidal solution in the dilute alkali. After aging to develop a more highly condensed gel with a higher average molecular weight, the gels are brought to the point of coagulation by the addition of sodium chloride and forced through a spinneret directly into a dilute aqueous sulfuric acid solution. The filament of coagulated xanthate decomposes in acid to a cellulose fiber. This is called viscose rayon and is a handsome silky fiber, but the tensile strength is not as great as that of natural silk. Viscose gels forced through narrow slits into the acid create films of cellulose known as cellophane. Viscose-derived materials are classed as reconstituted cellulose, and all suffer from some deficiencies in physical properties, such as wet strength. These deficiencies probably arise from unavoidable degradation of the cotton cellulose by the alkaline solution used in the process.

Partially acetylated cellulose is soluble in acetone and forms spinnable gels in that solvent. After spinning, the solvent can be removed leaving fibers of acetate rayon. Until the development of nylon in the third decade of this century, acetate rayon was the dominant synthetic fiber. Cellophane of acetate rayon is still a high-volume industrial commodity, and cellulose acetate sponges, produced by salting to full coagulation of a cellulose acetate gel, are superior in properties to natural sponges.

Although cellulose rayons have been largely displaced by other synthetic fibers and plastics, such as nylons, polyesters, and polyamides, there remains an interesting conversion of wood cellulose. The conversion of cotton or wood cellulose to cellulose fibers with a greater rigidity similar to linen fibers remains an economically attractive goal.

At such time as interest is rekindled in using this versatile, abundant, natural polymer, a superior nondegradative method of producing usable dispersions and gels of cellulose already exists.[11] This method, which uses concentrated aqueous calcium thiocyanate as a dispersion medium and various dilution and washing techniques for coagulation, results in cellulose materials with properties superior to existing technology. Improvements in coagulation techniques could provide the linen like polymer mentioned in the preceding paragraph.

CARBOHYDRATE SERIES

Dehydration of plant carbohydrates can occur in discrete steps with the loss of water intramolecularly or intermolecularly. The *Intramolecular* dehydration creates a double bond. *Intermolecular* dehydration is the condensation reaction to longer molecular chain lengths. Generally, however, conditions capable of dehydrating cellulose produce long-chain unsaturated polymers.

Related to these long-chain unsaturated polymers, the next lower series of carbohydrates are the polyketides or polyenols. The empirical formula for this series is $C_{2n}(H_2O)_n$. Structures for the two possible forms for these carbohydrates are:

Polyketide Unit **Polyenol Unit**

The polyenol is a most attractive structure, with conjugation of the π-electrons allowing for extensive delocalization. The anionic form would also be highly favored in aqueous alkaline solutions. The most popular assignment for the structure is the unconjugated ketonic structure. It is possible that this form is favored in certain solvents other than water. However, when treated with strong aqueous alkali, polyketones cyclize to phenols and polyphenols.[12] This reaction would probably be more facile for the anion of the polyenol form. It is probable that the phenols listed in Table 3.2 are produced from polyketide or polyenol intermediates.

The series of carbohydrates resulting from further general dehydration are the ketenes, $C_{2n}H_2O$. These are highly reactive towards many reagents and are avidly hydrophilic. Ketene, $H_2C=CO$, the simplest member of the series reacts vigorously with water to form acetic acid. The reverse of this reaction has not been demonstrated, although the reaction that led to its discovery utilized the pyrolysis of acetic anhydride.[13] Preparation directly from acetic acid should be possible using dessiccants more effective than ketene itself. The fact that this method of forming ketene has not been demonstrated possibly indicates that there are few molecules that react with water in such a favorably energetic manner. The current method is the pyrolysis of acetone.[14] The *anti*-addition of water to ketene to form glycolaldehyde, $HOCH_2CHO$, has also not been reported. With diazomethane ketene reacts cleanly to give cyclopropanone.[15]

The next homolog in this series, butatrienone, is apparently not a linear molecule.[16] It is bent with adjacent carbons orthogonal to one another. Hexatetrenone is predicted to have the same kind of structure. This implies some degree of radical character at each carbon. It is unlikely that a paired structure would gain sufficient bond energy, even invoking hyperconjugation, to overcome the pairing energies involved. The propadienone also has a bent structure. The odd-numbered ketenes are not members of this series of carbohydrates.

The members of this series of carbohydrates should possess an acidity greater than the odd-membered ketenes. This is a simple consequence of the nature of the anion, $HC \equiv C - C \equiv C - O^-$. Alternant resonance forms are not possible for the odd-numbered ketenes. It is also impossible to hydrate the odd-numbered ketenes to form polyhydroxyaldehydes or ketones, so they are not members of carbohydrate series. They are, however, carbohydrates in the literal use of the word.

ACETIC ACID CHEMISTRY

Acetic acid, on the other hand, is a carbohydrate both in linguistic usage and as a member of a series. Ketoacids and polyketoacids comprise the homologs of this series. A chemical confirmation of the carbohydrate classification for acetic acid is furnished when acetic acid vapor is brought into contact with phosphorus pentoxide. These conditions readily produce a caramel similar in all respects to those obtained when sugars are dehydrated by heat.

As an abundant consequence of the ubiquitous vinegar eel's appetite for alcohol, its waste product, acetic acid, is one of the most ancient of reagents. The alkali metal salts of acetic acid have been used to prepare acetone for nearly two hundred years. Boyle is credited with having first observed the liquid produced by heating potassium acetate. Perkin [17] describes the formation of acetone from a compound that he had reported years earlier.[18] The structure he proposed for this salt is:

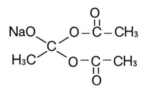

Sodium Orthoacetate

Perkin suggests the clean pyrolysis of this salt to acetone, sodium acetate, and carbon dioxide as the mechanism for acetone formation.

The yield of acetone from the pyrolysis of the acetate depends upon the metal salt used in the synthesis.[19] Calcium acetate decomposed in the presence of lime is the most effective salt for producing the highest yields. The generally accepted equations for this reaction,

$$Ca(OOCCH_3)_2 = H_3CCOCH_3 + CaCO_3 \qquad (1)$$

or

$$Ca(OOCCH_3)_2 = H_3CCOCH_3 + CaO + CO_2 \qquad (2)$$

suggest that this could easily be the consequence of the high crystal-lattice energy of calcium carbonate or calcium oxide.

The effect of metal ions upon the yields of acetone and some of the by-products of this reaction is summarized in Table 3.3.[19]

Other interesting reactions may account for the failure of the salts of the alkali metals in the acetone competition. They can also help to understand the underlying equilibria that rationalize the amounts of methane listed in Table 3.3. Fry[20] describes "the well-known reaction:"

$$H_3CCOONa + NaOH = CH_4 + Na_2CO_3 \qquad (3)$$

This reaction is an interesting example of the encryptation of valuable scientific knowledge. Many members of the previous generation of chemists must have performed this reaction in undergraduate laboratories as the preferred method of preparing small amounts of pure methane.

Table 3.3 Comparison of the Yields of Acetone from Different Metal Acetates

Metal Acetate	Percent of Theoretical Acetone	Percent of Volatile Oils	Methane as Percent of Total Gases
LiOOCCH$_3$	101.0	—	14.1
NaOOCCH$_3$	37.5	37.0	44.9
KOOCCH$_3$	11.6	52.0	32.5
Mg(OOCCH$_3$)$_2$	75.9	5.5	1.0
Ca(OOCCH$_3$)$_2$	82.5	5.0	30.1[a]
Ba(OOCCH$_3$)$_2$	89.0	10.0	15.0[a]

[a] The total gas produced in these experiments was only 10% of that produced by the decomposition of the alkali metal salts. In relative terms sodium acetate generated 1500% more methane than calcium acetate.

In the training of the present generation of chemists, knowledge of the reaction became obscure. A rapid informal survey of many colleagues across the country has failed to discover anyone aware of this reaction. It is perhaps understandable, in view of the many demands of modern pedagogy and the ever diminishing time available to the education of students, but it may also reflect a pervasive opinion that there is little to learn from the chemical literature produced by less highly instrumented research. It may also reflect opinion that we are capable of learning the truth of chemistry with our "modern" approach. Such attitudes ignore the obvious fact that the very best instrument for scientific research, the highly trained human intellect, has not changed appreciably in thousands of years.

The reaction in equation (2) implies an equilibrium that raises some interesting chemical possibilities in accounting for acetone formation. The equilibrium shown in equation (4):

$$H_3CCOO^- + 2OH^- = CH_3^- + CO_3^{-2} + H_2O \qquad (4)$$

may become operative at high activities of OH^- and low activities of water. The methylide ion as a possible intermediate in the formation of acetone should be considered. The electrostatics of the reaction appear to be unfavorable with an anion adding to another anion. This may not be as unlikely as first appearances suggest, however. Similar reactions of hydride ion are often invoked[21] for reductions occurring under alkaline conditions and suffering the same electrostatic disadvantage. An equation for an appropriate equilibrium for the methylide case is:

$$H_3COO^- + CH_3^- + H_2O = H_3COCH_3 + 2OH^- \qquad (5)$$

The effects of water and hydroxide upon this reaction are just the reverse of those for equation (4). Acetone and methane would be formed in inverse relative amounts depending upon these two reaction parameters. The formation of oils is a measure of the extent of aldol condensation characteristic of such alkaline conditions. The fact that barium acetate produces more oils than either magnesium or calcium acetate is interesting in view of the relatively small amount of methane formed from this salt. It suggests that carbon methylation by methylide ion might be responsible for the formation of some of the "oils" in the cases of sodium and potassium acetates. The observed facts, summarized in Table 3.3 can be rationalized in a satisfactory manner by this hypothesis.

It would have been useful to know the states of hydration of the various salts used in this work. The alkali metal acetates generally are more hygroscopic than the alkaline earth salts. This reaction must be quite sensitive to water activity. By-product yields should be decreased and condensation products should be increased with increasing water activity. Barium acetate, being the least hygroscopic of the salts listed, confirms this trend.

A difference between reagent grades of potassium hydroxide and sodium hydroxide that is sometimes overlooked is the fact that the potassium base contains 15% water while the sodium salt contains only 2%. Reactions using pure potassium hydroxide always have higher water activity than those carried out using sodium hydroxide.

This difference has been exploited in the optimization of the production of formate in the reaction of carbon monoxide with caustic. The pertinent equilibria are:

$$CO + OH^- = HCOO^- \tag{6}$$

$$CO + OH^- + H_2O = H_2 + HCO_3^- \tag{7}$$

$$H_2 + OH^- = H_2O + H^- \tag{8}$$

$$H^- + HCO_3^- = HCOO^- + OH^- \tag{9}$$

$$2HCOO^- = {}^-OOC-COO^- + H_2 \tag{10}$$

Mixtures of sodium and potassium hydroxide can be employed to adjust the water activity in these melts to favor any of these equilibria except those shown in equations (7) and (10).

Conversions of carbohydrates to reduced compounds of carbon *in vitro* are dependent upon various solution parameters such as the activities of water, hydroxide ion, and carbonate ion as well as the activities of the various forms of soluble carbohydrate ions. All of the chemistry discussed in this chapter has been carried out at atmospheric pressure or in autoclaves operating at some fixed relief pressure. In none of these cases can the reactant solutions be considered to be in true equilibrium with the gas phase. In Chapter IV, following, reactions in closed systems under autogenous pressure will be considered.

These reactions, functioning in closed systems, manifest behavior that can best be interpreted as being under equilibrium conditions. The effects of changing parameters upon the amounts and nature of the observed products in these closed-vessel reactions support the ideas that have been promulgated in this chapter.

REFERENCES

1. M. Calvin, *J. Chem. Educ.*, **26**, 639 (1949)
2. E. Fischer, *Ber. Dtsch. Chem. Ges.*, **23**, 2114 (1890); *ibid.*, **24**, 1836 (1891).
3. T. Funazukuri, R. R. Hudgins, and P. L. Silvaston, *Ind. Eng. Chem. Process Des. Dev.*, **25**, 172−181 (1986).
4. C. A. Lobry de Bruyn and W. Alberda van Ekenstein, *Recl. Trav. Chim. Pays-Bas*, **14**, 203 (1895).
5. J. M. de Bruijn, A. P. G. Keeboom, and H. van Bekkum, *Recl. Trav. Chim. Pays-Bas*, **105**, 176−183 (1986).
6. J. U. Nef, *Liebig's Ann.*, **357**, 1494 (1907), *ibid.*, **376**, 1 (1910).
7. W. L. Evans, *Chem. Rev.*, **31**, 537−560 (1942).
8. P. M. Molton, R. K. Miller, J. M. Donovan, and T. F. Demmitt, *Carbohyd. Res.*, **75**, 199−206 (1979).
9. H. R. Appell, in *Fuels from Waste*, L. L. Anderson D. A. Tillman, Eds., Academic Press, New York, 1977, Chap. 8.
10. C. F. Schönbein, *Archives des sciences physiques et naturelles* (1846)
11. M. F. Bechtold and J. H. Werntz, *U. S. Pat.*, #2,737,459 (1952) and #2,810,162 (1957).
12. U. Weiss and J. M. Edwards, *Aromatic Systems Derived Through the Polyketide Pathway* in *Biosynthesis of Aromatic Compounds from Acetic Acid*, Wiley/Interscience, New York 1980, Chap. 17.
13. N. T. M. Willsmore, *J. Chem. Soc.*, **91**, 1938 (1907).
14. J. Schmidlin and M. Bergman, *Ber. Dtsch. Chem. Ges.*, **43**, 2821−2823 (1910).
15. N. J. Turro and W. B. Hammond, *Tetrahedron*, **24**, 6017−6028 (1968).
16. L. Farnell and L. Radom, *J. Am. Chem. Soc.*, **106**, 25−28 (1984).

17.   W. H. Perkin, *J. Chem. Soc.*, **49**, 324–328 (1886).

18.   W. H. Perkin, *J. Chem. Soc.*, **31**, 185 (1868).

19.   W. Krönig, *Z. Angew. Chem.*, **37**, 667 (1924).

20.   H. S. Fry, E. L. Schulze, and H. Weitkamp, *J. Am. Chem. Soc.*, **46**, 2768–2275 (1924).

21.   C. G. Swain, A. L. Powell, T. J. Lynch, S. R. Alpha, and R. P. Dunlap, *J. Am. Chem. Soc.*, **101**, 3584–3587 (1979).

Chapter IV

Redox Equilibrium Reactions

THERMODYNAMICALLY FEASIBLE EQUILIBRIA

The treatment of various carbonaceous materials with KOH solutions of high ionic strength, *in vacuo*, results in the oxidation of the carbonaceous material to carbonate by any water present with the formation of hydrogen gas. In a closed tube these carbonaceous materials react with hydroxides and carbonates of sodium, potassium, magnesium, and calcium in redox–disproportionation reactions to form a variety of compounds with carbon in a reduced state. Aliphatic anions from acetate to hexanoate, alkanes from C_2H_6 to $C_{17}H_{36}$, and aromatics such as phenols, benzene, and benzoic acids are formed under different equilibrium conditions. These same solutions fix nitrogen. Gaseous nitrogen is chemically combined to form ammonia, pyridine, glycine, glutamic acid, and other carbon–nitrogen compounds. This fixing occurs at modest pressures and temperatures compared with the classic Haber process. The addition of other oxides, such as B_2O_3, Al_2O_3, and MnO_2, to the basic reaction mixture modifies the reaction so that the products obtained are alcohols, ethers, and olefins.

79

An elementary course in organic chemistry usually requires covering the sum of knowledge of various class reactions and fostering the ability to select reactions appropriate to a given sequence. The typical "road map" question of how to accomplish this is often the most frustrating of challenges. After a firm grounding in elementary chemistry a student longs to be able to write a balanced equation such as:

$$7C \ + \ 3H_2O \quad = \quad C_6H_5OH \quad + \quad CO_2 \tag{1}$$

Such a response could be viewed by an instructor as a specious answer undeserving of credit or comment. However, the equilibrium chemistry of carbon and water described in the preceding paragraph lends validity to such a response. These reactions proceed to yield product in amounts usually deemed acceptable for organic syntheses.

A thermodynamicist faced with the same equation would be impressed by the simplicity of the chemistry displayed but would question the thermodynamic feasibility of such a reaction. It is difficult, using the "numerology" of thermodynamics, to prove such feasibility, but it is a simple matter using the principles of thermodynamics to do so. Where reactions are known to occur they are conclusively proved to be feasible. Where several feasible reactions can be added to one another to give a sensible summary reaction, then this reaction also is feasible. A simple example can be used to illustrate this point.

When calcium oxide is heated in the presence of carbon at temperatures near its melting point, the acetylide CaC_2 is formed along with CO.[1] Treatment of the acetylide with water produces acetylene and the metal oxide. The overall reaction is a water-assisted redox-disproportionation of elemental carbon to acetylene and CO:

$$CaO + 3C = CaC_2 + CO \qquad (2)$$

$$CaC_2 + H_2O = HC\equiv CH + CaO \qquad (3)$$

The chemical conversion of acetylene to other organic molecules in simple reactions effected at modest temperatures is documented in the early literature. Of particular interest is the reaction of acetylene with a molten mixture of alkalies, at $220°C$, reported[2] to produce good yields of acetate.

Fry[3] examined this reaction at temperatures in excess of $300°C$. He was, perhaps, the first to speculate upon the displacement of hydride by action of alkali upon certain hydrocarbon compounds. Methane, some unsaturated hydrocarbons, hydrogen and elemental carbon are formed under these conditions. Reactions of carbon with water to form carboxylates, alkanes and alkenes are feasible under suitable conditions. Under alkaline conditions the following reactions readily occur:

$$HC\equiv CH + H_2O = H_3CCHO \qquad (4)$$

$$H_3CCHO + H_2O = H_2 + H_3CCOOH \qquad (5)$$

$$H_2 + CO = C + H_2O \qquad (6)$$

Equation (7) is the sum of equations (1) through (6):

$$2C \quad + \quad 2H_2O \quad = \quad H_3CCOOH \qquad (7)$$

Since all of the reactions listed are thermodynamically feasible, as written, it follows from the principle of additivity that the reaction described in equation (7) is feasible. Although it would be a longer sequence, equation (1) can be similarly proved to be thermodynamically possible.

PRESSURE DEPENDENT EQUILIBRIA OF CARBON

Early reports furnish abundant "aboriginal" evidence for a pressure-dependent equilibrium chemistry of carbon. This early work helps to understand the results just described. In the following discussion, pressure-dependent equilibria are operative. The earlier workers, here cited, reported these results without considering the equilibrium implications.

For example: When benzaldehyde is treated with concentrated KOH, it disproportionates to benzoic acid and benzyl alcohol[4]:

$$2C_6H_5CHO \quad + \quad 2OH^- = \quad C_6H_5COO^- + \quad C_6H_5CH_2O^- + \quad H_2O \quad (8)$$

When n-butanol is treated with concentrated KOH at "about $250°-275°C$ in an autoclave," butyrate and hydrogen are formed[5]:

$$CH_3CH_2CH_2CH_2OH \quad + \quad OH^- = \quad CH_3CH_2CH_2COO^- + 2H_2 \quad (9)$$

Equation (8) can be represented as the sum of three competing equilibria, equations (10), (11), and (12):

$$C_6H_5CHO \ + \ OH^- = C_6H_5CH(OH)O^- \tag{10}$$

$$C_6H_5CH(OH)O^- \ = C_6H_5COO^- \ + \ H_2 \tag{11}$$

$$H_2 \ + \ C_6H_5CH(OH)O^- \ = \ C_6H_5CH_2O^- \ + \ H_2O \tag{12}$$

For the benzaldehyde disproportionation to proceed as in equation (10), the equilibrium hydrogen pressure for equation (11) and equation (12) must be less than 1 atm (the product hydrogen of equation (11) is the reactant hydrogen of equation (12); one pressure is characteristic of both reactions). The reverse of equation (12) is readily demonstrated. When benzyl alcohol is treated with strong caustic, *in vacuo*, hydrogen is evolved.[6] In the butanol reaction the equilibrium between butanoxide and water shown in equation (9) requires the equilibrium pressure for hydrogen to be greater than the autoclave pressure at 275°C for the reaction to proceed as written. Equilibrium (13) requires that water be consumed as hydrogen is evolved. In this equilibrium butanoxide is accurately described as a drying agent.

$$CH_3CH_2CH_2CH_2O^- \ + \ H_2O \ = \ CH_3CH_2CH_2COO^- \ + \ 2H_2 \tag{13}$$

A careful study[7] of reaction (9) demonstrates how a shift in the various reactant activities affects product distribution. With caustic soda, hydrogen evolution ceased after 3 hr at the autoclave temperature of 275°C. Good yields of butyric acid, with smaller amounts of 2-ethylhexanoic acid, were obtained. When metallic sodium was used, resulting in lower water activity, equivalent amounts of butyrate and ethylhexanoate were produced; Equation (14) describes this equilibrium.

$$2CH_3CH_2CH_2CH_2OH + OH^- =$$
$$CH_3CH_2CH_2CH_2CH(C_2H_5)COO^- + 2H_2 + H_2O \quad (14)$$

The criterion for a disproportionation of the type illustrated in equation (8) is currently held to be a lack of hydrogen substitution on the α carbon of the aldehyde undergoing the Cannizzaro reaction. When such hydrogens are present, the aldehyde undergoes a reaction similar to the Dumas reaction of equation (9). Formaldehyde meets the criterion for the Cannizzaro reaction, yet when it is treated with aqueous caustic at elevated temperatures hydrogen is generated as in the Dumas reaction.[3,8] Also, when butyraldehyde is treated with aqueous caustic in a closed vessel, under autogenous pressure, little hydrogen is produced. These different reactions are apparently examples of similar pressure-dependent redox equilibrium reactions.

The crossed Cannizzaro reaction exploits the equilibrium:

$$H_2CO + OH^- = HCOO^- + H_2 \quad (15)$$

When an equimolar mixture of formaldehyde and benzaldehyde are treated with aqueous caustic, both possible acids and alcohols can be found among the products.[9] With a 30% excess of formaldehyde mainly formate and the phenolate are formed. These data are rational with an assumption that at moderate temperatures the equilibrium pressure for equation (15) is less than 1 atm and probably higher than the equilibrium pressure of the reaction in equation (12).

The presence of formate in these reaction solutions also introduces several other equilibria:

$$CO + OH^- = HCOO^- \tag{16}$$

$$CO + OH^- + H_2O = H_2 + HCO_3^- \tag{17}$$

$$H_2 + OH^- = H_2O + H^- \tag{18}$$

$$H^- + HCO_3^- = HCOO^- + OH^- \tag{19}$$

$$2HCOO^- = {}^-OOC-COO^- + H_2 \tag{20}$$

Le Châtelier's principle also requires the equilibrium pressures for hydrogen in these reactions to be the same as that in equation (15).

REDOX HYDROLYSIS OF CARBONACEOUS MATERIALS

Reactions of various carbonaceous materials with water, in the presence of oxides and carbonates of the alkali and alkaline earth metals, have been examined in closed systems. The equilibria described in equations (1) and (7) are observed directly in these experiments.

Carbonaceous materials as widely different as graphite, poly(tetrafluoroethylene), triirondodecacarbonyl, coal, lignite, sawdust, corncobs, crystalline cellulose, starch, sugar, and carbon monoxide, when mixed with reagent-grade potassium hydroxide (85% KOH, 15% H_2O) and the reaction mixture heated continuously in an evacuated glass vessel, produce hydrogen gas rapidly, as soon as the hydroxide melts (approximately $120°C$). Diamond dust does not react under this same set of conditions. The stoichiometry of the representative reaction is exactly that required by the equilibrium:

$$C + 2KOH + H_2O = 2H_2 + K_2CO_3 \qquad (21)$$

The reaction of dextrose is interesting in that it occurs in two stages. When the hydroxide first melts hydrogen is rapidly evolved until 2 mol of hydrogen per mole of dextrose are produced. The solution obtained from this treatment is a clear yellow liquid stable for extended periods of time, at temperatures below $200°C$, without further evolution of hydrogen. This experimental evidence demonstrates a reproducible endpoint in this redox reaction to produce a carbon compound with the average oxidation state of the cumulene salt shown below:

$$CH_2(OH)CH(OH)CH(OH)CH(OH)CH(OH)CHO + 4KOH =$$
$$H_2C=C=C=C=C(OK)_2 + 2H_2 + K_2CO_3 + 5H_2O \quad (22)$$

The product shown is in its dehydrated form. Other isomers include the other cumulene salt, $KOCH=C=C=C=CHOK$ and the alternate diyne salt, $KOH_2C-C\equiv C-C\equiv COK$. In this medium of high ionic strength the more polar isomer shown in the equation is the more likely.

The second stage ensues at $220°-350°C$ when the carbon is fully oxidized and the remainder of the hydrogen is evolved.

The ability of carbon dissolved in molten KOH to act as a powerful reducing agent has been long known. At low activities of water the potassium ion is reduced to potassium metal.[10,11]

When the reactants described above are combined under equilibrium conditions in sealed tubes, no product hydrogen is found. The evidence for these reactions being under equilibrium conditions is twofold:

When the sealed tube is rapidly (5 min) heated to $380°C$, the pressure in the tube increases to 65 atm and remains constant at that temperature. When the tube is cooled to room temperature as rapidly as possible, (15 min) the pressure inside the tube falls just as rapidly to a vacuum. The observed pressures are the same whether dextrose or graphite is used as the reactant. When dextrose is used elemental carbon is found at the end of the reaction. The observed pressure at $380°C$ is less than that of a KOH solution of comparable strength.

Variations in the quantities of reactants loaded into the tubes are reflected in the relative amounts of different products. These changes are most readily interpreted by the application of Le Châtelier's principle.

Under these equilibrium conditions redox—disproportionation reactions such as the following are observed:

$$2C + KOH + H_2O = CH_3COOK \qquad (23)$$

$$8C + 5 KOH + 3H_2O = CH_3CH_2CH_2CH_2CH_2COOK + 2K_2CO_3 \quad (24)$$

The pressures observed depend in part upon the amount of water present, the amount of carbonate present, and the temperature at which the tube is maintained during the reaction. With temperatures of $50°-400°C$ and water contents as high as 80% (by weight), these reactions display autogenous pressures of 3-67 atm. In some experiments gaseous hydrogen under pressure was added before the reaction tubes were sealed and heated. The effect of adding hydrogen to the reaction mixture is essentially that of carrying out the reactions with higher water content. The addition of active Raney nickel does not alter the effect of added hydrogen upon the reaction. Mixtures of hydrogen and carbon monoxide do not react, a matter that is of some interest in understanding the nature of this reaction, since carbon monoxide alone is a suitable carbonaceous reactant for this chemistry.[12] This is possibly a result of the equilibrium position of equation (20) lying well to the left as written.

For all reactions in concentrated KOH solutions with graphite, dextrose, or cellulose as the carbonaceous reactant, the carboxylates (acetate, propionate, butyrate, valerate, and caproate) were obtained. At low to intermediate water activities the relative amounts of each varies as equations (23) and (24) would predict. Lower water activities favor low-molecular-weight carboxylates while intermediate water activities favor higher-molecular-weight carboxylates. As the water content of the reaction mixture increases, lower carboxylates are again favored. In some cases heptanoate is found among the reaction products. The examples of highest water activity studied are 80% water (by weight). These solutions are still at very high ionic strength.

The hydroxides of sodium, potassium, magnesium, and calcium, as well as the carbonates of the same metals, are equally capable as reactants in this chemistry.

When carbonates are used, the oxidation product is bicarbonate and the reduction products are the saturated hydrocarbons, $C_{11}H_{24}$ through $C_{17}H_{36}$, along with some acetate.[12] Le Châtelier's principle predicts such an effect from the set of equations that includes (17) and (18). At higher activities of carbonate new equilibria are apparently established with the alkanes. A general expression for these is:

$$C_nH_{2n+1}COONa \ + \ H_2O \ = \ C_nH_{2n+2} \ + \ NaHCO_3 \qquad (25)$$

As the available hydroxide decreases, the carbon dioxide produced in the disproportionation reaction accumulates as bicarbonate ion. The hydrogen ion activity increases and the carboxylates remain in solution to undergo decarboxylation as illustrated.

The addition of other oxides to the reaction mixtures changes the nature of the products. Boric oxide, B_2O_3, and aluminum oxide, Al_2O_3, shift the major products from carboxylates to alcohols, esters, and ethers. Tetrahydrofuran and *tert*-butyl alcohol are typical products obtained when the alkaline reaction mixture is acidified. The autogenous pressures observed in these experiments are higher than the typical reaction in molten potassium hydroxide.

The role of these oxides has not been studied in detail, but they are likely to establish new equilibria that adjust the water activity. The products are those that are typical of the Meerwein–Ponndorf–Verley reaction,[13,14] the Oppenauer oxidation,[15] or the Tishchenko reaction.[16] This class of reactions constitutes a variation of the Dumas reaction described in equations (9) and (13). The formation of partially hydrated alkoxyborates and aluminates should effectively lower the activity of the water in the reaction mixture. While there has been a large volume of work on the detailed mechanism for the Meerwein–Ponndorf–Verley reaction, the reaction mechanisms all assume that the aluminates are soluble species. It is yet more likely that soluble aluminates are stable in these solutions. They are at considerably higher ionic strengths than those used in the typical Meerwein–Ponndorf–Verley reaction. For the Meerwein–Ponndorf–Verley reaction to be invoked here requires the conversion of carboxylates, known from the unmodified experiments to be major products, to be converted to alkoxides. Reductions involving hydrogen, hydride, or formate such as

$$(H_3CCOO)_4Al^- + 8H_2 = (H_3CCH_2O)_4Al^- + 4H_2O \qquad (26)$$

probably occur. The aluminum oxide or boric oxide can prevent the precipitation of the carboxylates allowing these new equilibria to operate.

The acidification of the product aluminum alkoxides can produce chemistry beyond that accomplished in the autogenous pressure vessel. Depending upon the activity of water present in this procedure and the acidity of the ultimate reaction solution, alcohols, esters, or ethers can be formed.

The tertiary butanol produced here raises interesting possibilities in this chemistry. Fry[3] describes an interesting reaction whereby acetate in strongly alkaline solutions generates methane. This reaction has not attracted a great deal of discussion although for a time it was used as a standard laboratory preparation for pure methane.

In the previous chapter this example of the encryptation of knowledge was noted. Many among the preceding generation of chemists must have prepared methane using this reaction. They chose not to transmit this knowledge to currently active chemists. It is difficult to find a contemporary chemist who knows of this reaction. A balanced equation for this preparation is:

$$H_3CCOO^- + OH^- = CH_4 + CO_3^{2-} \qquad (27)$$

It is possible to demonstrate the reverse of equation (27) whereby acetate is formed in molten KOH at $350^{\circ}C$ under a high pressure of methane.[12] The formation of acetate is facilitated by the addition of carbonate to the molten KOH as Le Châtelier's principle demands. These facts suggest other equilibria that will help in understanding the formation of *tert*-butanol in the aluminum oxide modified redox–disproportionation reactions currently being discussed. Equilibria involving methylide ion are possible in this case. The pertinent equations are:

$$CH_3COO^- + 2OH^- = CH_3^- + CO_3^{2-} + H_2O \qquad (28)$$

$$CH_3COO^- + CH_3^- = CH_3C(CH_3)(O^-)_2 \qquad (29)$$

$$CH_3C(CH_3)(O^-)_2 + CH_3^- + H_2O = (CH_3)_3CO^- + 2OH^- \quad (30)$$

Other evidence for these equilibria will be discussed when the equilibrium model is discussed in detail.

The products formed when manganese dioxide is added are yet more selective. This modification apparently produces only methanol and 2-butene.[12] This highly selective result is understandable as a consequence of known chemistry of manganese compounds. It does not require invoking any special transition metal catalysis, as one might be tempted to do.

It is a corollary of inorganic chemistry that, for an element capable of forming compounds in a range of oxidation states, alkaline solutions favor the extreme oxidation states and acidic solutions favor intermediate oxidation states. When chlorine is dissolved in aqueous potassium hydroxide and the solution heated, potassium chloride and potassium chlorate are the stable chlorine compounds. It is probable that, were potassium chlorate not insoluble under these conditions, potassium perchlorate would be the oxidized product of this disproportionation. When solid chlorate is heated, it in turn disproportionates to chloride and perchlorate. Conversely, when perchlorate solution, to which an excess of chloride has been added, is acidified, chlorine is generated. Strongly oxidizing agents, such as potassium permanganate or bismuthate, would be difficult to obtain chemically were it not for this characteristic redox−disproportionation chemistry.

Manganese dioxide is an intermediate oxidation state for manganese. When it is placed in strongly alkaline conditions, as in the reaction now being discussed, it will disproportionate to Mn^{2+} and either MnO_4^{2-} or MnO_4^-. These are both capable of oxidizing acetate to methanol in alkaline solution as in equation (31):

$2MnO_4^- + 5H_3CCOO^- + 4OH^- =$

$$5H_3CO^- + 2Mn^{2+} + 5CO_3^{2-} + 2H_2O \quad (31)$$

giving a sensible provenance for one of the products of this highly selective reaction.

The other product, 2-butene, is apparently the analog of tetrahydrofuran in the reactions where aluminum oxide is added. It is likely that the tetrahydrofuran in that case is produced during the acidification of the cyclic aluminate:

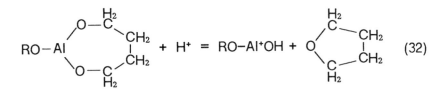

$$(32)$$

For Mn^{2+} the analogous cyclic ether and its likely behavior in acidic solution are given by:

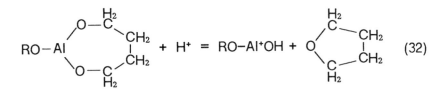

$$(33)$$

The equilibrium model for the reactions under autogenous pressure must account for the constant pressure, observed during the reaction at temperature, and the rapid rise and fall of pressure with temperature. In the overall reaction cycle no gaseous products are allowed. The pressure behavior is the same whichever solid carbonaceous reactant is employed. This independence of reactant supports the assumption that the reactions for different carbonaceous materials are identical or closely similar for all cases where other parameters are constant. Gaseous carbonaceous reactants introduce complexities that will be discussed later.

The pressure behavior demonstrates virtually instantaneous establishment of stacked equilibrium reactions that connect a solid reactant (graphite) with a series of products (carboxylate salts for the unmodified graphite–KOH reaction) with different solubilities in the reactant solution. The resulting phase equilibria at both ends of the chain essentially *clamp* the positions for the stack of intermediate reactions. The various components of the gas phase are in equilibrium with the entire stack of reactions. The participation of the gaseous components in the equilibrium stack is proven by the fact that at the end of the reaction cycle, after the reactor is cooled to room temperature, the pressure in the reactor vessel is identical to that before the heating cycle was initiated. Gaseous reactants constitute a different equilibrium situation since a given gaseous reactant will affect only those equilibria in which it can participate. Only an exact mixture of gaseous reactants, matching the mixture autogenously produced during the heating of the reaction mixture, could be added without displacing the equilibrium position.

The composition of the vapor phase at reaction temperature is not easy to determine with any accuracy. Removal of an analyzable fraction of the vapor phase during the reaction dramatically alters the equilibrium position in any small-scale reaction. This result is as the Le Châtelier principle predicts. Introduction of an inert gas to the equilibrating system also perturbs the position of the system, lowering the activity of all gaseous reactants and affecting only those reactions in the stack that have a gas as a reactant or product. It does serve, however, to trap gaseous components to allow identification and the estimation of their *relative* activities.

It is a corollary of this model that when the reactant initially produces carbon species in solution, as in the case of all carbohydrates, very little of the initial reactant carbon precipitates as elemental carbon or graphite. In most examples only minor amounts of the carbohydrate carbon remains as graphite after 8 hr at the clamped equilibrium position. Apparently, major amounts of product carboxylates precipitate from the reaction mixture before any elemental carbon is formed. The amount of each carboxylate precipitated depends upon various reaction parameters.

The constant autogenous pressure observed in these reactions reflects the constancy of all reactant descendants present in the liquid phase. The same reactant descendants apparently are obtained whether graphite, coal, or a carbohydrate is the carbonaceous reactant. The rate of the reaction where graphite is the reactant is dependent upon the solubility product constant for graphite in equilibrium with the reaction mixture compared with the solubility product constants for the roster of products. The comparison of various carbonaceous reactants and their products is given in Table 4.1.

Table 4.1 Comparison of Different Carbonaceous Reactants[a]

Reactants Consumed			Products				
Carbon	KOH	CO_3^{2-}	CH_3COOH	C_2H_5COOH	C_3H_7COOH	C_4H_9COOH	$C_5H_{11}COOH$
G- 37	19	8	6.6	1.08	0.73	1.18	0.344
D-115	53	2.3	44.5	4.0	0.223	0.031	0.031
C- 9.3	6.6	1.5	3.25	0.23	0.012	0.037	0.065

Other Reaction Parameters

Carbon Source	Initial $C/H_2O/KOH$	Carbon Consumed	Carbon Balance
(G)raphite	260/118/80	14%	90%
(D)extrose	132/243/254	87%	95%
(C)oal	24.5/56/33.9	38%	100%

[a]All quantities in millimoles except as noted

The reactions that produced the data for Table 4.1 were carried out in a reactor different from the apparatus used for the remaining experiments in this chapter. The reactor was shaken and there were grinding/stirring balls inside the reactor instead of an externally actuated stirrer. All of the tubing connecting the reactor to the control valves and pressure gauge was quarter-inch diameter as opposed to the one-sixteenth-inch diameter tubing, used throughout the microreactor system. It also was possible for those experiments summarized in Table 4.1 to establish a water piston at the pressure gauge before the reactor was plunged into the thermostatted fluidized-bed sand bath. The fluidized-bed sand bath also was much larger for this series of experiments. The pressure and equilibrium behavior of these reactions was examined in this larger-scale apparatus.

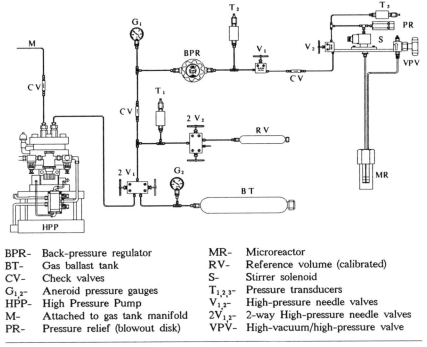

BPR-	Back-pressure regulator	MR-	Microreactor
BT-	Gas ballast tank	RV-	Reference volume (calibrated)
CV-	Check valves	S-	Stirrer solenoid
$G_{1,2}$-	Aneroid pressure gauges	$T_{1,2,3}$-	Pressure transducers
HPP-	High Pressure Pump	$V_{1,2}$-	High-pressure needle valves
M-	Attached to gas tank manifold	$2V_{1,2}$-	2-way High-pressure needle valves
PR-	Pressure relief (blowout disk)	VPV-	High-vacuum/high-pressure valve

Figure 4.1 High-Pressure Volumetric System and Microreactor

VOLUMETRIC APPARATUS FOR HIGH-PRESSURE STUDIES

The high-pressure experiments that generated the data in this chapter were mainly performed in an apparatus as shown in Fig. 4.1. The reaction vessel, a microreactor fabricated entirely of Hastelloy C^{TM} with a nominal volume of 5 ml, was stirred by a magnetically actuated plunger designed to produce vigorous stirring. After being charged with solid and liquid reactants, the reactor was evacuated and connected to a high-pressure manifold fitted with an adjustable back-pressure regulator to maintain the reactor at a chosen constant pressure.

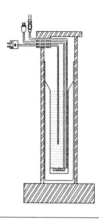

Figure 4.2 Sand Bath Assembly

The gases were supplied to the back pressure regulator from a reference volume (approximately 15 ml volume) containing an initial pressure well above the regulated pressure. The amount of gas consumed in the reaction was determined from the pressure drop in the reference volume, assuming ideal gas behavior. Both the reactor and the reference volume were fitted with precision pressure transducers (Model TH-2VA 0-5000 psi, T-Hydronics, Inc., Columbus, Ohio) allowing accurate measurement of the pressure at any stage of a given reaction.

If the pressure in the reference volume fell to values close to the regulated pressure maintained in the reaction vessel, the reference volume was repressurized from a large ballast tank (nominal volume 150 ml). The ballast tank was pressurized from a standard tank gas manifold by means of an air-operated, single-ended, diaphragm compressor (Model J46-14205-1,10,000 psi, Superpressure, Inc., Silver Spring, Md.) at the outset of an experiment.

The stirred reaction vessel was heated in a thermostatted fluidized-bed sand bath, Fig. 4.2. This was enclosed in a cylindrical explosion barrier (10 cm diameter, 50 cm length) capable of withstanding an explosion of 5 g of TNT without rupture. All components of the system were hydraulically tested to a pressure of 5000 psig.

Where rapid heating was desired for an experiment, the sand bath was heated to the required temperature before the assembly was raised to enclose the reactor.

For each experiment the volumes of the reference volume as well as the charged reactor were determined by a procedure using argon gas over a wide range of pressures. This served to test the validity of the assumption of ideality as well as furnishing a correction for the volume occupied by the nongaseous reactants.

When the reaction came to equilibrium, with no further gas being consumed, it was discontinued and the reactor closed and removed from the high-pressure system and heating bath. At this point the gases remaining in the supernatant volume of the reactor were removed to a vacuum apparatus. Using standard vacuum-line techniques these gases were analyzed. This analysis furnished a precision check on the volumes as determined in the high-pressure apparatus. Agreement of the data from the high-pressure experiment and the vacuum-line analysis was within 1%.

The solid reaction products remaining in the reactor were analyzed in an appropriate fashion.

For reactions carried out under autogenous pressure the reactor vessel was charged with reactants, calibrated with argon gas as previously described, evacuated, and closed. The closed reactor then was immersed in the thermostatted sand bath, and the pressures in the reactor were monitored by means of the pressure transducer. For these reactions all of the tubing external to the heated zone had to be filled with water to serve as a pressure-transmitting piston to the transducer. If this is not done, water distills away from the heated zone and it is impossible to estimate the reactant water with any precision.

SUPERIORITY OF CELLULOSE AS A REACTANT

From Table 4.1 it can be seen that the reactions are virtually identical in the roster of products. The main differences arise from the initial behavior of the carbonaceous reactant. Graphite and coal behave similarly. The higher percentage conversion shown for the coal used is an artefact of the initial amount. The total amount of reaction in each of these two cases was essentially the same. This is a persuasive argument for the similarity of these reactants in their equilibrium with the solution. One difference between the two solids is the carbon balance, with the carboxylates accounting for 90% of the carbon consumed for graphite and 100% for the coal. As we will see from more complete product analyses, these differences can be readily rationalized. The incomplete material balance for graphite is probably the consequence of phenol formation in this case. These experiments did not include quantitative analysis for the phenols.

Another deduction to be drawn from Table 4.1 is that conversions are considerably higher for dextrose. When the reactant initially produces carbon species in solution (as in the case of dextrose) very little of the initial carbon survives to precipitate as elemental carbon. In the experiment described in Table 4.1 only 13% of the dextrose carbon remains as elemental carbon after 8 hr at the "clamped" equilibrium position. Apparently, in the latter case, to reach equilibrium major amounts of product carboxylates precipitate from the reaction mixture before any graphite is formed.

The carbohydrates are the reagents of choice for comparative reactions since the amount of product is maximized in the initial establishment of equilibrium. Carbohydrates dissolve rapidly to form solutions of active carbon. This allows the assessment of the effect of changing activities of carbon upon the roster of products in a set of reactions. For solid reactants, such as graphite, the carbon activity in the reactant medium is fixed by the solubility product for a given solution. The carbon activity remains virtually constant at constant temperature. Altering the temperature can adjust the carbon activity, but this is only one of many parameters that will be so affected.

Reactions of carbohydrates will be used to illustrate the effects of various parameters upon the roster of products. More product allows a more complete and accurate analysis. The carbohydrate of choice is microcrystalline cellulose powder. It is a good standard reagent, since it is available in a highly pure state and is not hygroscopic.

PROBLEMS IN USING THE MICROREACTOR

With the large-scale apparatus (used to obtain the results in Table 4.1), it is virtually impossible to obtain the precision of quantitative data possible in the microreactor system (used to generate the data shown in all subsequent tables in this chapter). There are, however, several systematic problems encountered in using the microreactor. The problem of filling the pressure transducer's water piston external to the heated reactor system has already been described. Another difficulty results from the limited heat capacity of the smaller sand bath relative to the heat capacity of the reactor, stirrer, and connector tubing. When the microreactor system is employed, the reaction mixture does not reach the desired reaction temperature until after approximately one hour. The 15-sec temperature rise and the accompanying rapid pressure rise typical for the larger-scale shaker experiments are not possible for the microreactor. The grinding action of the stirring balls in the shaker-type experiment also cannot be reproduced in the micro reactor system.

The slower temperature rise of the microreactor compared with the larger-scale apparatus has a special consequence when carbohydrates are used as reactants. The low-temperature reaction that results in the formation of the highly reducing carbon solutions has been described in equation (22). When the temperature rise is slow, this reaction occurs before any disproportionation can take place. The hydrogen evolved in this low-temperature reaction would, in the larger-scale reactions, be reabsorbed by the equilibrium reaction solution when a sufficiently high temperature was reached to produce the onset of the redox–disproportionation reaction. In the smaller tubing of the microreactor this is diffusion limited and is subject to a range of factors that effectively prevent the ready reabsorption of the hydrogen produced in the low-temperature degradation of the cellulose. In some reactions most of the hydrogen is reabsorbed. In others it is not. For those cases where it is not, the carbon consumed by this reaction must be subtracted from the total initial carbon in order to assess the carbon balance for the disproportionation reaction. In all of the tables that follow, the initial carbon has been corrected, using equation (21), for the amount of the hydrogen remaining at the end of the reaction.

When graphite is used as the carbonaceous reactant in the microreactor system, it cannot be activated by grinding. Activation by carbon monoxide reduction is required for this system. So for reactions of graphite, the amount of carbon monoxide used for activation must be added to the amount of initial carbon reactant. The graphite reactions described hereafter have all been corrected for this factor. The percentage of the initial carbon that forms products, other than elemental carbon, is given in each table as percent conversion. The portion of initial carbon undergoing the disproportionation reaction that is accounted for by the analytical procedures is shown as the material balance percentage.

REACTIONS OF CELLULOSE WITH WATER AND ALKALI METAL COMPOUNDS

Since all of the reactants are participants in the equilibrium stack, the relative activities of water and hydroxide, along with the total availability of the carbon in solution, furnished as carbohydrate, determine the product options available for a given set of these parameters.

The roster of products obtained from the reactions of cellulose is identical to that obtained when graphite is the carbonaceous reactant. Relative amounts of individual compounds will vary, however, since cellulose produces higher activities of reactive carbon in solution than the solubility of graphite allows. Some products found in trace amounts in reactions of graphite are produced in measurable quantities when cellulose is used as the carbonaceous reactant. All indications point to the same set of mechanistically important equilibria for both reactants.

Water is a reagent and a component of the solvent for the disproportionation. At low activities it limits the extent of the redox reaction by restricting the hydrogen and oxygen available to the equilibrium stack. At low activities of water the activity of hydroxide must be high and fairly constant in the reactant solution. The effects of this set of parameters are apparent in Table 4.2. Acetic acid, requiring only hydration of the carbon, is dominant among the carboxylate products. Phenols and substituted phenols are also products when the reaction is so limited and the availability of dissolved carbon relatively high.

Under the conditions described for Table 4.2 there is so little water in the reactant solution that it is best described as molten potassium hydroxide diluted by water. Several equilibria are probable in this medium that help in understanding the product distribution for this reaction. While the equations given are not intended as an exhaustive list, they serve to illustrate points of comparison with reactions under conditions of higher water activities to be discussed later.

Table 4.2 Complete Carbon Balance for Cellulose in Molten Potassium Hydroxide[a]

Reagents: (g/mmol)[b]
85% KOH = 2.31/35.0, H_2O = 0.35/18.9; $C_6(H_2O)_5$ = 0.51/19.0

Products	Yield[c]	
	mol%	wt%
CH_3COOH	40.7	31.2
C_2H_5COOH	27.3	25.8
C_3H_7COOH	4.5	5.1
C_4H_9COOH	4.1	5.4
$C_5H_{11}COOH$	1.8	2.7
C_6H_5OH	6.4	7.7
$CH_3C_6H_4OH$	8.3	11.5
$C_2H_5C_6H_4OH$	6.8	10.6

[a] Low Water Activity (Under 60 psig of Argon)
[b] CO_2 accounts for 1.29 mmol of initial carbon, reduced products for 7.82 mmol.
[c] Conversion = 62.7%; Material Balance = 98.9%.

$$2C + OH^- = HC\equiv CO^- \qquad (34)$$

$$HC\equiv CO^- + 2OH^- + 2H_2O = HCOO^- + CO_3^{2-} + 3H_2 \qquad (35)$$

$$HCOO^- + 2OH^- = H^- + CO_3^{2-} + H_2O \qquad (36)$$

$$H_2 + OH^- = H_3O^- \qquad (37)$$

$$H_3O^- = H^- + H_2O \qquad (38)$$

The reaction in equation (37) has been observed in the gas phase.[17] The hypervalent ion, H_3O^-, has been described as *long lived* in these ion-cyclotron resonance mass spectrometry experiments. Its equilibrium constant with water and hydride ion, shown as equation (38) has been estimated [18] and the equilibrium, as written, lies far to the right. Both of these equilibria help to explain the anomalous increase in the solubility of hydrogen gas in melts of potassium hydroxide with low water activities.[19] With mixtures in which water activities are below that for the monohydrate of potassium hydroxide, hydrogen solubilities increase with decreasing water content. Above the monohydrate composition hydrogen solubilities decrease with decreasing water content. The equilibria shown in equations (37) and (38) can also account for the equivalence of hydrogen gas and water in the chemistry of the carbon redox disproportionations currently under discussion.

Equation (36) is an expression that describes formate as a source of hydride and is seen to be favored in this role under conditions of low water activity, another simple application of Le Châtelier's principle. The absence of any formate ion among the products listed in Table 4.2 also suggests its consumption in effective reductions by hydride. Formate is found among the products if the disproportionation reaction is carried out at temperatures below $250^{\circ}C$ with low water activity. Reactions under high pressure involving carbon monoxide and potassium hydroxide, where formate is usually a product under atmospheric pressure, will be discussed fully. In these reactions, also, formate is found among the final carboxylate products.

One possible source of formate in equation (36) is the alkaline degradation of ketene anion in these molten potassium hydroxide solutions, as shown in equation (35). The ketene arises from the alkaline degradation of dehydrated cellulose or from the disso-lution of solid carbon, as shown in equation (34). Ketene anion also probably plays a critical role in producing the longer chain carboxylates in those reactions in which graphite is the carbonaceous reactant. At appropriate activities condensations such as the following can occur.

$$2HC{\equiv}CO^- = HC{\equiv}C{-}C{\equiv}CO^- + OH^- \qquad (39)$$

The subsequent hydrogenation and hydration of the product of equation (39) would produce butyrate ion. All even numbered carboxylates can be similarly produced. Odd numbered carboxylates would have to arise from degradation of even numbered ketene anions in a manner similar to the reaction shown in equation (35).

Increasing the water activity in these equilibrium reactions has the predicted effects upon the yields of the various carboxylate ions. The distribution of products for a reaction carried out with water at moderate levels of activity is given in Table 4.3.

Table 4.3 Complete Carbon Balance for Cellulose in Molten Potassium Hydroxide[a]

Reagents: $(g/mmol)^b$
85% KOH = 2.32/35.0, H_2O = 1.36/57.5; $C_6(H_2O)_5$ = 0.50/18.6

Products	Yield[c]	
	mol%	wt%
CH_3COOH	36.0	30.1
C_2H_5COOH	52.2	53.8
C_3H_7COOH	4.8	5.9
C_4H_9COOH	4.1	5.4
$C_5H_{11}COOH$	0.6	1.0
C_6H_5OH	0.9	1.1
$CH_3C_6H_4OH$	1.1	1.7
$C_2H_5C_6H_4OH$	0.3	0.6

[a] Moderate Water Activity (Under 200 psig of Argon)
[b] CO_2 accounts for 1.27 mmol of initial carbon, reduced products for 9.82 mmol.
[c] Conversion = 61.7%; Material Balance = 97.3%.

At these intermediate levels of water content more of the redox reaction can occur, and propionic acid represents the major carboxylate product. The amount of phenolic products also decreases dramatically, as Table 4.3 indicates. As the amount of water is increased to the point where it is in excess of the monohydrate of potassium hydroxide, the equilibria of equations (37) and (38) are both displaced well to the left. The consequence of this is a dominance of hydrogen as the principle reducing agent in all of the descendant equilibria of the stack. The reaction medium in this range is best described as an aqueous solution of high ionic strength. In this medium, conditions for condensations should be more favorable.

In this region of activities just above the composition of the potassium hydroxide monohydrate, the relative amounts of the carboxylate product change in favor of the higher-molecular-weight products and the phenolates disappear completely. It is possible to carry out reactions in this range of water activities where caproate is the major product and where heptanoate is also produced.

As the ionic strength of the reactant medium decreases with the increase of water content, a secondary effect is observed. The chain length of the alkyl carboxylate constituting the major product again decreases. Condensation reactions, important in determining the average chain length of the solution species undergoing hydrogenation, apparently are becoming less favorable. It is also possible that in this range of water activities degradation reactions to produce carbonate and hydrogen continue to be reasonably efficient. With more hydrogen available for hydrogenation, the lower carboxylates would again be favored in the redox reaction.

The relative quantities of potassium hydroxide and available carbon *in solution* also have a determining effect upon this redox disproportionation reaction. If the amount of hydroxide is such that a major portion is consumed in the reaction, the reaction medium gradually changes from hydroxide to carbonate. As the carbonate activity increases, alkanes begin to appear among the reduced products. Methane can also be detected in the gas phase during the reaction along with carbon dioxide. Carbon dioxide in the gas phase demonstrates the emergence of the equilibrium:

$$CO_3^{2-} + H_2O + CO_2 = 2HCO_3^- \qquad (40)$$

The reaction medium is being buffered to lower pH values when this equilibrium is operative. The main effect is to change the carboxylate equilibria to that shown in its general form as equation (25). The extreme of this condition can be examined by using sodium carbonate as the alkaline reactant.

Table 4.4 Complete Carbon Balance for Cellulose in Sodium Carbonate (High Base-to-Cellulose Ratio)[a]

Reagents: (g/mmol)[b] $Na_2CO_3 = 0.543/5.126$, $H_2O = 0.6(est)/33.3$; $C_6(H_2O)_5 = 0.148/5.491$		
Products	Yield[c]	
	mol%	wt%
$C_{11}H_{24}$	11.0	10.0
$C_{12}H_{26}$	61.5	60.6
$C_{13}H_{28}$	23.6	25.1
$C_{14}H_{30}$	3.8	4.3

[a] Low Water Activity
[b] CO_2 accounts for 1.82 mmol of initial carbon, reduced products for 3.29 mmol.
[c] Conversion = 81.4%; Material Balance = 98.2%.

The experiment summarized in Table 4.4 is at relatively low water activity. Increasing the water available to the equilibrium stack increases the chain length of the saturated hydrocarbons produced. Mixtures of saturated hydrocarbons, $C_{11}H_{24}$ to $C_{20}H_{42}$, have been obtained from this simple reaction.

The alkanes are all linear hydrocarbons and no olefins are produced under these conditions. This point will be addressed in detailed discussions of the pertinent equilibria. It is probably a consequence of the Varrentrap reaction.[20] This reaction converts oleate, $H_3C(CH_2)_7CH=CH(CH_2)_7COO^-$, to acetate, H_3CCOO^-, and palmitate, $H_3C(CH_2)_{14}COO^-$, by treatment with concentrated aqueous sodium hydroxide. In the course of the reaction the double bond of oleate is shifted into conjugation with the carbonyl, where it is more susceptible to alkaline cleavage. In solutions of high ionic strength branched chain ions would probably rearrange to the linear isomer in order to minimize electrostatic energies.

It is possible by varying reactant ratios to obtain different rosters of products. The products obtained in Table 4.4 are those obtained when the ratio of base to reactant cellulose is high. With lesser amounts of available base the products listed in Table 4.5 are found.

This product roster is particularly informative since it includes carboxylate products characteristic of the reaction in potassium hydroxide melts. The various furans and pyrans observed in this reaction are obviously formed in the potassium hydroxide reaction as intermediates.

Table 4.5 Complete Carbon Balance for Cellulose in Sodium Carbonate (Intermediate Base-to-Cellulose Ratio)[a]

Reagents: $(g/mmol)^b$ $Na_2CO_3 = 1.12/10.56$, $H_2O = 1.3(est)/72.3$; $C_6(H_2O)_5 = 0.60/22.25$		
Products	Yield[c]	
	mol%	wt%
CH_3COOH	27.5	19.4
C_2H_5COOH	2.7	2.4
C_3H_5COOH	1.5	1.6
C_8H_{14}	2.0	2.6
C_9H_{16}	6.4	9.3
$C_{10}H_{18}$	10.2	16.5
$C_{11}H_{20}$	2.4	4.3
$C_4H_6O_2(\gamma\text{-lactone})$	29.8	30.1
$C_5H_8O_2(\text{methyl-}\gamma\text{-lactone})$	3.6	4.2
$C_5H_{10}O(\text{methyltetrahydrofuran})$	9.5	9.6
(Unidentified)	4.4	—

[a] Moderate Water Activity
[b] CO_2 accounts for 2.93 mmol of initial carbon, reduced products for 12.81 mmol.
[c] Conversion = 72.3%; Material Balance = 95.6%.

With an adequate supply of base relative to the amount of carbon in solution these intermediates are converted into the products listed in Table 4.2. In the present example they remain as observable products since all of the base is consumed before the supply of carbon in solution is exhausted.

It is also apparent from the data given in Table 4.5 that the amount of available base limits all of those equilibria, previously discussed, that generate the hydrogen necessary for all reductions. The fact that the linear hydrocarbons listed for this reaction are lower homologs than those listed in Table 4.4 is a consequence of this limitation.

Table 4.6 Complete Carbon Balance for Cellulose in Sodium Carbonate (Low Base-to-Cellulose Ratio)^a

Reagents: (g/mmol)^b		
$Na_2CO_3 = 0.54/5.08$, $H_2O = 1.2(est)/66.7$; $C_6(H_2O)_5 = 1.20/44.40$		
Products	Yield^c	
	mol%	wt%
CH_3COOH	44.3	38.5
C_2H_5COOH	10.5	11.3
C_3H_7COOH	3.4	4.3
C_4H_9COOH	0.6	0.9
C_4H_8	0.8	0.7
C_5H_{10}	1.4	1.4
C_6H_{12}	2.1	2.6
C_7H_{14}	4.2	6.0
$C_4H_6O_2$(γ-lactone)	6.5	8.1
$C_5H_{10}O$(methyltetrahydrofuran)	21.1	26.3
(Unidentified)	5.1	—

^{a} Moderate Water Activity

^{b} CO_2 accounts for 3.26 mmol of initial carbon, reduced products for 21.86 mmol.

^{c} Conversion = 66.8%; Material Balance = 94.8%.

With even less available base, where the ratio of carbonate base to reactant cellulose is very low, the mix of products confirms these conclusions, as seen in Table 4.6. The furans and pyrans observed as products in this experiment, as previously pointed out, also demonstrate this limitation.

The conditions for the reaction summarized in Table 4.6 must closely approximate those for the reaction summarized in Table 3.2 since the products are similar. Unless the unidentified products are phenolics, we see none in this case.

Again, the consequences of the limited hydrogen available for reduction is manifest by the even shorter chain length of the product hydrocarbons. In this example another consequence of the limited base is the dominance of furans and pyrans among the products.

This set of experiments give an enlightening picture of the various classes of reactions that dominate the equilibrium stack. It is beyond the scope of the present discussion to list all of the probable sets of equilibria that participate in the equilibrium stack. The complexity of the interconnections among the various sets of equilibria are made apparent in this particular set. It will require many more experiments to establish the detailed stack for any given set of reaction parameters.

REACTIONS OF CELLULOSE WITH WATER AND ALKALINE EARTH COMPOUNDS

The use of magnesium or calcium hydroxide as the active base affords another approach in unscrambling this mess. Both of these bases can be used to operate the redox reactions under fairly constant hydroxide ion concentration. The total available base to carbon in solution can be set at satisfactorily high levels without appreciable effect upon the hydroxide ion activity in the reaction medium. Solid magnesium or calcium oxides can be added in appropriate amounts for hydrolysis to metal ions and hydroxide ion at essentially constant activity to achieve this result. The consumption of water in the redox reaction should occur without great changes in hydroxide ion activity in such a system.

This reaction gives information about the coal-forming reaction. The discussion of the dependence of redox reactions upon pH earlier in this chapter states the general rule for an element with multiple available oxidation states: in basic medium the extremes of oxidation states are the main result, and in acidic medium, intermediate or elemental oxidation states result. This generalization leaves unanswered the question of how acidic or how basic is the medium required in each case.

It is difficult to estimate the pH of the reactant medium for the experiment summarized in Table 4.7. It is possible to state that, for this reaction, the hydroxide ion activity does not vary greatly as the carbonaceous reactant is consumed. It is also possible to say that the pH of this reactant mixture is higher than that for the reaction carried out in equilibrium with magnesium oxide (Table 4.8).

From these experiments, it is clear that, at activities of the hydroxide ion below that of molten potassium hydroxide solutions, the solubilities of the various products are increased. As a consequence, more elemental carbon is precipitated as equilibrium is attained. The formation of small quantities of methane and ethane in these reactions is similarly significant in coal formation. At yet lower pH elemental carbon is the major product.[21] For the reaction in $Mg(OH)_2$ it is a significant product, with other reduced products comprising the remainder.

The first observation of interest in comparing the reactions of the two alkaline earth oxides is the change in the amount of graphite deposited. The difference between the observed conversion and total conversion is the amount of elemental carbon found at the end of the reaction.

Table 4.7 Complete Carbon Balance for Cellulose in Equilibrium with Calcium Hydroxide[a]

Reagents: (g/mmol)[b]		
Ca(OH)$_2$ = 1.0/13.5; H$_2$O = 7.0(added)/389; C$_6$(H$_2$O)$_5$ = 0.51/18.8		
Products	Yield[c]	
	mol%	wt%
Carboxylic acids:		
CH$_3$COOH	3.20	1.93
CH$_3$CH$_2$COOH	6.80	5.07
CH$_3$CH$_2$CH$_2$COOH	6.50	5.76
CH$_3$CH$_2$CH(CH$_3$)COOH	9.96	10.23
CH$_3$CH$_2$CH$_2$CH$_2$COOH	2.00	2.34
CH$_3$CH$_2$CH$_2$CH(CH$_3$)COOH	3.00	3.50
CH$_3$(CH$_2$)$_4$COOH	1.04	1.36
CH$_3$(CH$_2$)$_2$CH(CH$_3$)COOH	1.58	2.07
CH$_3$(CH$_2$)$_5$COOH	0.80	1.16
Dienes:		
C$_6$H$_{10}$	3.20	2.64
C$_7$H$_{12}$	2.12	2.05
C$_8$H$_{14}$	5.80	6.43
C$_9$H$_{16}$	16.44	20.54
C$_{10}$H$_{18}$	1.00	1.39
Alkenes:		
C$_7$H$_{14}$	3.12	3.09
C$_8$H$_{16}$	1.50	1.69
Hydroxyl Alkenes:		
C$_5$H$_{10}$O	5.46	4.73
C$_6$H$_{12}$O	6.74	6.79
Alchohols:		
CH$_3$CH(OH)CH$_3$	3.38	2.04
(CH$_3$)$_2$CHCH$_2$OH	3.08	2.30
Furans:		
2-propyl-furan	1.40	1.55
Alkyl Phenols:		
C$_8$H$_{10}$O	1.40	1.72
C$_9$H$_{12}$O	3.60	4.94
C$_{10}$H$_{14}$O	3.10	4.68
(Unidentified)	3.80	—

[a] Moderate Water Activity (under 33 psig of argon).
[b] CO$_2$ accounts for 4.20 mmol of initial carbon, reduced products for 11.08 mmol.
[c] Conversion = 81.7%; Material Balance = 96.2%.

Table 4.8 Complete Carbon Balance for Cellulose in Equilibrium with Magnesium Hydroxide[a]

Reagents: (g/mmol)[b]		
$Mg(OH)_2$ = 1.0/13.5; H_2O = 1.0(added)/55.6; $C_6(H_2O)_5$ = 0.51/18.9		

Products	Yield[c]	
	mol%	wt%
CH_3COOH	5.03	2.87
C_3H_7COOH	3.21	2.68
C_4H_9COOH	4.90	4.75
$C_5H_{11}COOH$	3.78	4.16
C_8H_{14}(Diene)	9.79	10.22
C_9H_{16}(Diene)	4.20	4.93
$C_{10}H_{18}$(Diene)	9.93	12.88
$C_4H_8O_3$(Trioxepane)	13.87	13.69
5-methyl-2-furanone	2.52	2.34
$CH_3C_6H_4OH$	5.59	5.73
$C_2H_5C_6H_4OH$	3.80	4.40
$C_3H_7C_6H_4OH$	16.50	21.30
HOC_6H_4OH	9.64	10.06
(Unidentified)	7.27	—

[a] Low Water Activity (under 30 psig of argon).
[b] CO_2 accounts for 1.99 mmol of initial carbon, reduced products for 7.75 mmol.
[c] Conversion = 55.8%; Material Balance = 92.7%.

The amount of carbon found increases in a sensible fashion as the pH of the reaction medium moves to lower values. In potassium hydroxide reactions very little is found. For the series of reactions discussed in this chapter, magnesium hydroxide is the least efficient in producing reduced compounds and the most efficient in producing elemental carbon. If this trend is followed to its ultimate conclusion, the conversion of cellulose to coal should be the dominant process in acidic oxides. This chemistry suggests that hydrocarbon formation and coal formation proceed from the same starting materials. Which is formed depends mainly upon the pH of the geological environment. In an environment at high pH the redox disproportionation dominates. At low pH dehydration is the main result.

The complexity of the product roster apparently increases with decreasing hydroxide ion activity. The calcium hydroxide reaction produces more different products than the potassium hydroxide reaction. The magnesium hydroxide reaction is also prolific of diverse products. It is clear from an examination of the various rosters of products that it is the redox–disproportionation reaction that is diminished under these circumstances. Phenols and substituted phenols are prevalent among the products, with increasing quantities of these with falling pH. These are found as products of the potassium hydroxide reaction only under water-starved conditions. For water starved reactions in potassium hydroxide, however, dehydration products like unsaturated furans and pyrans are not produced. These appear to be favored in the alkaline earth oxide reactions, in spite of reasonably high water activities in the initial reactant mixture. Acetate is found, as in most of the disproportionation reactions we have examined. Again, neither formate nor oxalate is detected. The furans and pyrans are possibly produced from cyclic ionic precursors, to be discussed, during the acid workup of the reaction mass.

These furans and pyrans as well as the hydrocarbons were found as products of the sodium carbonate reactions summarized in Tables 4.5 and 4.6, with intermediate and low base to cellulose ratios. They also are typical products in autoclave reactions, summarized in Table 3.2 on page 48, in which the base to cellulose ratios are very low, but where an excess of water vapor is furnished at elevated temperature.

This paradoxical roster of products is readily understood. In the water-starved reactions, such as that summarized in Table 4.2, the reactant phase is essentially molten potassium hydroxide in the region of anomalous hydrogen solubility.[19] In this solvent the activities of water, hydrogen, and hydride ion are tied by the equilibria described in equations (35) through (38). The amount of available water thus limits the total amount of reduction possible. The average oxidation state of the entire roster of reduced products is necessarily closer to that of the reactant cellulose. Precipitation of phenolate salts instead of carboxylate salts to establish the equilibrium clamp is the accommodation of the system to this limitation.

Oxygenated products, such as furans and pyrans, are the consequence of base-limited disproportionation, in which the pH is falling as the equilibrium is established. This falling pH also affects the solubility of the carboxylate salts. The carboxylate salts remain in solution at concentrations and in a pH range where decarboxylations to alkanes readily occur. The low solubility of alkanes in the reaction phase produces a new phase equilibrium between a hydrocarbon liquid phase and the reaction medium. In these reactions solubility, declining pH, and the amount of redox disproportionation act in concert to produce the rosters of Tables 4.5 and 4.6.

The reactions, with calcium oxide or magnesium oxide as the available base, are in the pH range where decarboxylations readily occur so the alkanes in the roster are expected. For the phenolic and oxygenated products to coexist it is necessary that the solubilities of the alkaline earth phenolates and alkaline earth cyclic diolates and enediolates differ significantly from the corresponding potassium or sodium salts. These differences in solubility suggest early precipitation during the establishment of equilibrium. Calcium acetate enjoys moderate water solubility while calcium phenoxide is much less soluble. Specific solubilities of diolates and enediolates such as those illustrated in Fig. 4.3 have not been reported. There is a general rule of solubility for dipositive ions, however, predicting that salts with dinegative ions are considerably less soluble than their salts with uninegative ions.

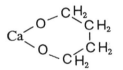

Calcium Butane-1,4-diolate

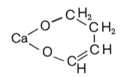

Calcium 1-Butene-1,4-diolate

Figure 4.3 Typical Alkaline Earth Cyclic Salts with Diols and Enediols

The cyclic structure given in Fig. 4.3 is not intended as a probable structure for the solid salts of these diolates. It is likely that in the crystalline salts a diolate ion is coordinated to different calcium ions. A representation of such a three-dimensional structure, while possibly more accurate, would comprise a confusing array. The natural bond angles of saturated and unsaturated carbon chains would obey similar stability rules in the three-dimensional or cyclic examples.

Salts of this type, along with the various phenolates, apparently precipitate to establish the solubility equilibria that clamp a different set of dominant stack reactions for the alkaline earth case.   Many more experiments are required before this set of reactions can be unscrambled.   The absence of aryl ethers and quinones from the product rosters of both of the alkaline earth reactions rules out any pyrolysis of the alkaline earth phenolates at the reaction temperature.

The presence of cyclopentenones among the products of the alkaline earth reactions lends some support to the model that emerges from these experiments.   Calcium 1,3-butadiene-1,4-diolate salts and their alkyl derivatives must also precipitate when the final equilibrium is established.   The reactions in which such salts are formed and the products to be expected from the acidification of these salts during the analytical workup will be discussed in detail later.

EFFECTS OF ADDED OXIDES ON THE REDOX DISPROPORTIONATION

Oxides can be added to modify the course of the redox disproportionation reaction in molten potassium hydroxide.   In the experiments where the effects of these oxide adducts were examined, graphite was used as the carbonaceous reactant.   To afford a more direct assessment of the role of these adducts, a typical reaction of graphite in molten potassium hydroxide is summarized in Table 4.9.

The addition of manganese dioxide to the redox–disproportionation reaction in molten potassium hydroxide also demonstrates the effect of a different kind of dipositive ion in these reactions at high ionic strength. Manganous ion differs from the alkaline earth ions, just discussed, in being amphoteric. In the strongly alkaline medium of these reactions it is present mainly as the manganite ion, $Mn(OH)_4^{2-}$. and the manganate ion, MnO_4^{2-}. The products of this reaction are presented in Table 4.10 and comprise the most selective product roster among the reactions studied.

Table 4.9 Complete Carbon Balance for Graphite in Molten Potassium Hydroxide[a]

Reagents: (g/mmol)[b]
85% KOH = 2.31/35.0; H_2O = 1.00/55.6; $C_{(gr)}$ = 0.52/43.3

Products	Yield[c]	
	mol%	wt%
CH_3COOH	7.7	4.4
H_3CCH_2COOH	3.0	2.1
$H_3CCH_2CH_2COOH$	6.0	5.0
$H_3CCH_2CH_2CH_2COOH$	23.8	23.0
$CH_3CH_2CH_2CH_2CH_2COOH)$	53.6	58.4
$CH_3CH_2CH_2CH_2CH_2CH_2COOH$	6.0	7.3
(Unidentified)	2.9	—

[a] Moderate Water Activity.
[b] CO_2 accounts for 1.65 mmol of initial carbon, reduced products for 5.31 mmol.
[c] Conversion = 16.6%; Material Balance = 97.1%.

Table 4.10 Complete Carbon Balance for Graphite in Molten Potassium Hydroxide with Manganese Dioxide[a]

Reagents: (g/mmol)[b]
85% KOH = 2.24/34.9; MnO_2 = 1.00/10.3 H_2O = 1.00/55.6; $C_{(gr)}$ = 0.51/42.6

Products	Yield[c]	
	mol%	wt%
CH_3OH	13.6	8.3
$H_3CCH=CHCH_3$	86.4	91.7

[a] Moderate Water Activity.
[b] CO_2 accounts for 1.66 mmol of initial carbon, reduced products for 6.39 mmol.
[c] Conversion = 19.0%; Material Balance = 99.7%.

The absence of the ubiquitous acetate and the presence of methoxide in its place can be rationalized by the following reaction:

$$MnO_4^{2-} + 2H_3CCOO^- + 4OH^- =$$
$$2H_3CO^- + 2CO_3^{2-} + Mn(OH)_4^{2-} \quad (41)$$

The manganite ion is the prevalent manganese species during the establishment of the final equilibrium position. The manganite cyclic ether that would give 2-butene upon acidification has been discussed in the introductory portion of this chapter.

Aluminum oxide also is amphoteric but, unlike the transition metals, *does not usually* display a range of oxidation states. Yet we see from the data in Table 4.11 that methanolate and 2-butene also are products of this reaction.

Table 4.11 Complete Carbon Balance for Graphite in Molten Potassium Hydroxide with Aluminum Hydroxide[a]

Reagents: (g/mmol)[b]		
85% KOH = 2.31/35.0; Al_2O_3 = 1.00/1.96; H_2O = 1.00/55.6; $C_{(gr)}$ = 0.52/43.3		

Products	Yield[c]	
	mol%	wt%
CH_3OH	9.1	4.2
$H_3CCH(OH)CH_3$	11.0	10.0
$H_3CCH=CHCH_3$	22.7	19.2
$(H_3C)_3COH$	6.1	6.8
$(CH_2)_4O(THF)$	18.3	19.9
$C_4H_9OH(?isomer)$	17.1	19.1
$CH_3OC(CH_3)_3$	7.8	10.4
$H_3CCH(OCH_3)CH_2CH_3$	6.1	8.1
$H_3COCH_2CH=CHCH_3$	1.7	2.2

[a] Moderate Water Activity.

[b] CO_2 accounts for 2.80 mmol of initial carbon, reduced products for 6.52 mmol.

[c] Conversion = 21.6%; Material Balance = 99.7%.

It is the presence of several other products in the roster of this reaction that demonstrates different sets of pertinent reactions for the aluminum and manganese adduct systems. Boric oxide as an adduct serves the same purpose as aluminum oxide. This similarity suggests a chemistry for this set of reactions similar to the chemistry of boranes so elegantly exploited by H. C. Brown.[22] The analogous chemistry of aluminum alkyls and the application of this chemistry to the present system will be discussed in detail later.

COMPARISON OF GRAPHITE AND CELLULOSE AS CARBONACEOUS REACTANTS

All of the data we have discussed to this point supports the assumption that the equilibrium stack for this system is essentially independent of the carbonaceous reactant. Speculation concerning the reactions comprising this stack must be reasonable for all carbon sources.

The equilibria involved must account for changes in the roster of products resulting from changes in reaction parameters. These limitations of causal commonalty reduce the possibilities of the mechanistically important equilibria to a manageable problem.

Earlier in this chapter, the reaction that solubilized cellulose at approximately $120°C$ was described. The dehydrated dinegative ion that was proposed, $H_2C=C=C=C=C(OK)_2$, apparently disproportionates further at temperatures in excess of $200°C$. The initial reaction at elevated temperatures establishes the partial equilibrium stack in the following equations:

$$H_2C=C=C=C=C(OK)_2 \; + \; OH^- \; + \; H_2O \; =$$
$$HC\equiv C-C\equiv CO^- \; + \; 2H_2 \; + \; K_2CO_3 \quad (42)$$

$$HC\equiv C-C\equiv CO^- \; + \; OH^- \; = \; 2HC\equiv CO^- \quad (43)$$

$$HC\equiv C-C\equiv CO^- + HC\equiv CO^- = HC\equiv C-C\equiv C-C\equiv CO^- + OH^- \quad (44)$$

The analogous reactions established when solid active graphite is the reactant are:

$$2C_{(gr)} \ + \ OH^- \ = \ HC{\equiv}CO^- \qquad (45)$$

$$4C_{(gr)} \ + \ OH^- \ = \ HC{\equiv}C{-}C{\equiv}CO^- \qquad (46)$$

$$6C_{(gr)} \ + \ OH^- \ = \ HC{\equiv}C{-}C{\equiv}C{-}C{\equiv}CO^- \qquad (47)$$

For both of these stacks we see that lowering of the hydroxide ion activity has the effect of increasing the chain length of the prevalent ketene(polyketene) anion. In the range of concentrations where the solvent is best described as molten potassium hydroxide, the addition of water to the reaction has this effect.

The solubility equilibrium reactions of graphite in molten potassium hydroxide represented in the usual equation format in equations (45) through (47) are perhaps more understandable if they are presented in the structural form employed in Chapter II. For equations (48) through (50) we have used a model reaction structure for an active graphite produced by the chemical reduction of typical graphite by carbon monoxide as in Chapter II (equations 8 and 9). Tribochemically activated graphite would present 1,3– biradical edges similar to the horizontal edges of the model structures used in the following equations.

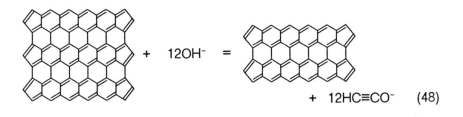

+ 12HC≡CO⁻   (48)

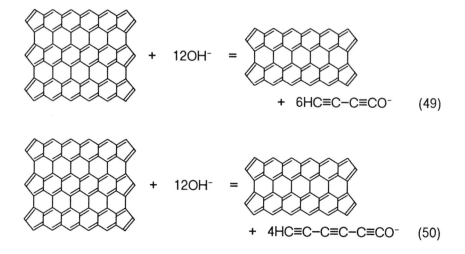

$$+ \quad 12OH^- \quad = $$

$$+ \quad 6HC{\equiv}C{-}C{\equiv}CO^- \qquad (49)$$

$$+ \quad 12OH^- \quad = $$

$$+ \quad 4HC{\equiv}C{-}C{\equiv}C{-}C{\equiv}CO^- \quad (50)$$

Without making any assumptions about the spin coupling in the graphite structure, all of these reactions are seen to be "spin-allowed". This set of equilibrium equations demonstrates the direct effect of hydroxide ion activity upon the the chain length of the prevalent ketene(polyketene) anion in solution. The stripping of the 1,3- biradical edges in units dependent upon such activity also is seen to be reasonable.

As the activity of water increases in these solutions, reactions in which it is a participant are affected. Typical primary reactions of the ketene(polyketene) anions with water are presented in the following equations:

$$HC{\equiv}CO^- + H_2O = H_3CCOO^- \qquad (51)$$

$$HC{\equiv}C{-}C{\equiv}CO^- + H_2O = H_2C{=}C{=}CH{-}COO^- \qquad (52)$$

$$HC{\equiv}C{-}C{\equiv}C{-}C{\equiv}CO^- + H_2O = H_2C{=}C{=}C{=}C{=}CH{-}COO^- \qquad (53)$$

Secondary reactions of the cumulene carboxylate anions produced in the primary hydrolyses can involve water, hydroxide ion, or molecular hydrogen as in the set of equations shown here for the four-carbon case:

$$H_2C=C=CH-COO^- + H_2O = H_2C=CH-CH(OH)COO^- \quad (54)$$

$$H_2C=C=CH-COO^- + 2OH^- + H_2O =$$
$$H_2C=CH-COO^- + CO_3^{2-} + 2H_2 \quad (55)$$

or

$$H_2C=C=CH-COO^- + 2OH^- + H_2O =$$
$$H_3C-CH_2-COO^- + CO_3^{-2} + H_2 \quad (56)$$

$$H_2C=C=CH-COO^- + H_2 = H_2C=CH-CH_2-COO^- \quad (57)$$

or

$$H_2C=C=CH-COO^- + 2H_2 = H_3C-CH_2-CH_2-COO^- \quad (58)$$

One of the most intriguing aspects of these high-pressure reactions is the higher relative yields of even-numbered carboxylates as opposed to odd-numbered carboxylates. The polyketene model nicely accommodates this fact, since the odd-numbered carboxylates arise chiefly from secondary degradative reactions. Hydrogen generated in these degradative reactions increases the relative importance of reactions like those represented by equations (57) and (58). Until there is sufficient water activity in the reactant medium none of these reactions can remove the longer-chain cumulene carboxylates from the reaction. Under water-starved conditions these precipitate from solution as graphite, phenols, or substituted phenols.

The magnesium and calcium diolate salts, the acidification of which would lead to the rosters of products in Tables 4.7 and

4.8, can arise in a sensible fashion from solutions of ketene(poly-ketene) anions, just described. The hydrogenation and cyclization steps are not necessarily in the sequence presented. Hydrogenation may be complete before the cyclization occurs.

The differences between this chemistry and that of the redox–disproportionation in molten alkali metal hydroxides is a consequence of the additional charge on the alkaline earth ions.

$$Ca^{2+} + {}^-OC{\equiv}C-C{\equiv}CH = (CaOC{\equiv}C-C{\equiv}CH)^+ \qquad (59)$$

Reaction of this positive ion with hydroxide ion would be the reverse of the reaction with a ketene anion. The product of this abnormal addition is a cumulene zwitterion salt:

$$(CaOC{\equiv}C-C{\equiv}CH)^+ + OH^- = Ca^+OCH{=}C{=}C{=}CH(O^-) \qquad (60)$$

The first hydrogenation step of this zwitterion salt would allow ready cyclization to an unsaturated diolate:

$$Ca^+OCH{=}C{=}C{=}CH(O^-) + H_2 = \overline{CaOCH{=}CH-CH{=}CHO} \qquad (61)$$

Further hydrogenation to the tetramethylene diolate salt could occur concurrently or sequentially:

$$\overline{CaOCH{=}CH-CH{=}CHO} + 2H_2 = \overline{CaOCH_2CH_2CH_2CH_2O} \qquad (62)$$

All of the examples presented are for the four-carbon poly-ketene anion. For the six- and eight-carbon anions similar sequences of reaction would produce various methyl- and ethyl-substituted diolate salts, both saturated and unsaturated. Acid hydrolysis of the saturated salts would produce the methyl- and ethylfurans and -pyrans seen in the product rosters. Acidification of the unsaturated salts would account for the various cyclo-pentanones and cyclopentenones observed. At hydroxide ion activities typical of the alkaline earth oxides the longer-chain polyketene anions should predominate.

CYCLIC ALUMINATES

Earlier in this chapter we invoked cyclic aluminates, in equation (32), and cyclic manganites, in equation (33), to account for differences in the product rosters observed when these metal oxides were added to the high-pressure disproportionation reactions in molten potassium hydroxide. It is useful here to discuss the manner in which such cyclic ethers can be formed.

For the aluminates, as with the dipositive alkaline earth ions, the simplest approach is to assume a reaction of the aluminate with a polyketene anion:

$$Al(OH)_4^- + HC\equiv C-C\equiv CO^- = [(HO)_3AlOC\equiv C-C\equiv CH]^- + OH^- \quad (63)$$

The product aluminate of equation (63) will hydrate in a manner different from the free polyketene anion. The probable course of this hydration is:

$$[(HO)_3AlOC\equiv C-C\equiv CH]^- \ + \ H_2O \ =$$
$$[(HO)_3AlOCH=C=C=CH(OH)]^- \quad (64)$$

Hydrogenation of the resulting cumulene aluminate can then occur as follows:

$$[(HO)_3AlOCH=C=C=CH(OH)]^- \ + \ 2H_2 \ =$$
$$[(HO)_3AlOCH_2CH_2CH_2CH_2(OH)]^- \quad (65)$$

In this solution cyclization by dehydration can readily occur:

$$[(HO)_3AlOCH_2CH_2CH_2CH_2OH]^- \ =$$
$$H_2O \ + \ [(HO)_2AlOCH_2CH_2CH_2CH_2O]^- \quad (66)$$

Reduction by hydride ion of the aluminate reactant in equation (66) also can occur in competition with the cyclization reaction:

$$[(HO)_3AlOCH_2CH_2CH_2CH_2OH]^- \ + \ H^- \ =$$
$$OH^- \ + \ [(HO)_3AlOCH_2CH_2CH_2CH_3]^- \quad (67)$$

The reaction of equation (67) would be favored over that of equation (66) as the activity of hydroxide ion decreases and the activity of the water increases. During the establishment of the final equilibrium position, such changes naturally occur as base is consumed. The ratio of hydroxide to initial carbonaceous reactant would control the final position for this competition.

When n-butoxy aluminate is heated in a concentrated solution of sodium butoxide rearrangements to more compact butoxyaluminates occur. The sec-butoxy and tert-butoxy aluminates can be detected in the equilibrated solutions. Such a sequence can account for the formation of the secondary and tertiary alcohols obtained from these high pressure disproportionations with added aluminum oxide. The ether products are probably formed during the acidification of the mixed aluminates on workup.

The behavior of aluminum *n*-butoxide, just described could account for the *sec*-butanol and the *tert*-butanol listed in Table 4.11, but the methanol, 2-propanol, and the ethers of these alcohols are difficult to rationalize in this fashion. Also, it would require much higher temperatures than those encountered during this reaction for the pyrolysis of aluminum butoxide to butenes. The observed 2-butene would be the least likely product of such pyrolysis in any event. It is probably significant that there is no acetic acid produced in this reaction.

A more satisfactory approach to this problem is to proceed from currently known chemistry of aluminum compounds that could form the roster of products. In all of these cases specific compounds of aluminum are chosen such that acid hydrolysis might be expected to yield every one of the products observed. In later discussions the mechanisms proposed for the formation of these aluminum compounds will be presented.

The minimum set of appropriate reactions is given in equations (68) through (71). This in not an exclusive list, but represents the best fit to our current model.

$$Al(OCH_3)_3 \ + \ 3H_2O \ = \ Al(OH)_3 \ + \ 3H_3COH \tag{68}$$

$$Al[OCH(CH_3)_2]_3 \ + \ 3H_2O \ = \ Al(OH)_3 \ + \ 3H_3CCH(OH)CH_3 \tag{69}$$

$$Al[OC(CH_3)_3]_3 \ + \ 3H_2O \ = \ Al(OH)_3 \ + \ 3(H_3C)_3COH \tag{70}$$

$$\overline{HOAl-CH_2-CH=CH-CH_2} + 2H_2O =$$
$$Al(OH)_3 + H_3CCH=CHCH_3 \quad (71)$$

The mixed ethers would result from the acid hydrolysis of mixed compounds, such as the reaction in equation (72). The absence of symmetrical ethers indicates that all of the aluminum compounds formed in arriving at equilibrium in this system are such mixed compounds. This is to be expected from a highly labile system.

$$Al(OCH_3)[OCH(CH_3)_2]_2 + 2H_2O =$$
$$Al(OH)_3 + H_3CCH(OH)CH_3 + H_3COCH(CH_3)_2 \quad (72)$$

It is necessary now to suggest a plausible mechanism, consistent with the conditions prevailing in this reaction, for the formation of aluminum compounds that are the reactants in equations (68) through (71).

The roster of products indicate carbon methylation as a dominant process that accounts for the preponderance of products. In this medium of molten potassium hydroxide the reaction described in equation (28) could furnish methylide ion for this purpose:

$$CH_3COO^- + 2OH^- = CH_3^- + CO_3^{2-} + H_2O \quad (28)$$

In the unmodified reaction there is too low a concentration of methylide to detect during the acidification procedure. The addition of boric oxide or aluminum oxide apparently introduces new equilibria that lower the methylide in equation (29) and remove it to the point that the acetate is lowered to undetectable levels at final equilibrium.

EXAMPLE OF BORANE CHEMISTRY

Since adding boric oxide to the reaction has an effect identical to that of aluminum oxide it is possible to invoke borane chemistry that has been described by Brown.[22]

The addition of carbon monoxide to organoboranes proceeds according to:

$$BR'R''R''' + CO = OBCR'R''R''' \tag{73}$$

A variety of oxidizing agents will oxidize the product of equation (73) to produce a tertiary carbinol in which the substituents are the same set as those in the original reactant organoborane.

$$OBCR'R''R''' + H_2O_2 + H_2O = B(OH)_3 + R'R''R'''COH \tag{74}$$

The mechanism for this remarkable reaction must account for the fact that the final product carbinol possesses the same set of substituent moieties as the reactant borane. Intermolecular reactions that would produce this result are improbable if not impossible. Brown suggests the following four steps as features for a satisfactory mechanism.

$$BR_3 + CO = R_3B^-C^+O \tag{75}$$

The product of equation (75) is analogous to borane carbonyl, H_3BCO, a volatile coordination compound formed readily when diborane and carbon monoxide are mixed. Apparently, in solution, this complex undergoes a methyl migration to form an acyl borane:

$$R_3B^-C^+O \ = \ R_2BCOR \tag{76}$$

A second alkyl migration produces a novel cyclic molecule:

$$R_2BCOR \ = \ R\overset{\displaystyle O}{\overset{\diagdown}{B-C}}R_2 \tag{77}$$

These three steps are plausible, displaying normal coordination for both carbon and boron. The sum of the first two steps is recognizable as a carbene insertion reaction with carbon monoxide behaving as a carbene. Such behavior for carbon monoxide is rare, but is observed in the case of electron-deficient molecules like the organoboron compounds and transition metal complexes. It may be that this insertion occurs in a single step. The third step (equation 77) is similar to many such migrations observed for transition metal complexes.

It is the final step of the proposed mechanism that is not evident:

$$R\overset{\displaystyle O}{\overset{\diagdown}{B-C}}R_2 \ = \ OBCR_3 \tag{78}$$

The product here is not coordinatively saturated, and this fact makes it difficult to understand a driving force for this reaction. The problem, however, is in the equation, not the chemistry. A more rational representation of this reaction is:

$$3RB{-}CR_2 \;=\; R_3CB\overset{O-B\overset{CR_3}{\diagdown}O}{\underset{O-B\diagup_{CR_3}}{\diagup}} \tag{79}$$

The product boroxole in equation (79) is pseudoaromatic and enjoys considerable resonance stabilization. Boroxoles are quite stable under the conditions of Brown's experiments. The resonance stabilization of the boroxole is adequate to serve as a driving force for the final migration in this chemistry.

Boroxoles are not stable under alkaline conditions like those of our reactions. In alkaline media the final step of this mechanism will necessarily be different, since the resonance stabilization of boroxole formation is denied as a driving force for this final step. Instead, the intermediate product of equation (77) will behave as in the following equation:

$$RB{-}CR_2 \;+\; OH^- \;+\; 2H_2O \;=\; [(HO)_3BOCR_3]^- \;+\; H_2 \tag{80}$$

Simple acid hydrolysis of the product of equation (80) produces the tertiary carbinol. Water has served as an oxidizing agent under the conditions of this reaction.

The consumption of acetate to furnish methylide ion already has been described. A reaction that removes methylide from solution to form intermediates similar to those in Brown's organoborane chemistry is required. It would be useful at the same time to account for the formation of methoxide as one of the products of such a reaction. The reactions described in equations (81) through (84) satisfy these requirements. In this alkaline medium, intermediates for the organoborane chemistry and the analogous aluminum alkyl chemistry will exist in their ionic forms.

With acetate readily available the following steps are suggested for the reaction where aluminum oxide is an adduct:

$$Al(OH)_4^- + {}^-OOCCH_3 = [(HO)_3AlOCOCH_3]^- + OH^- \quad (81)$$

$$[(HO)_3AlOCOCH_3]^- + CH_3^- =$$
$$[(CH_3O)(OH)_2AlCOCH_3]^- + OH^- \quad (82)$$

$$[(CH_3O)(OH)_2AlCOCH_3]^- + CH_3^- =$$
$$[(CH_3O)(OH)Al{-}\overset{\displaystyle O}{C}(CH_3)_2]^- + OH^- \quad (83)$$

$$[(CH_3O)(OH)Al{-}\overset{\displaystyle O}{C}(CH_3)_2]^- + CH_3^- + 2H_2O =$$
$$[(CH_3O)(OH)_2AlOC(CH_3)_3]^- + OH^- + H_2 \quad (84)$$

While equation (84) may offend a sense of chemical propriety, with methylide and water coexisting in the same equation, the water activity must be very low. It is perhaps less jarring to present an equation that will serve to summarize this reaction in which acetate is consumed and methoxide, *tert*-butoxide, and hydrogen are formed. The hydrogen arises from the role of water as the oxidizing agent in this chemistry.

$$[Al(OCOCH_3)_4]^- + 6OH^- = [H_3CO(OH)_2Al(OC(CH_3)_3]^-$$
$$+ 3CO_3^{2-} + 2H_2 + H_2O \quad (85)$$

Equation (85) is a combination of equations (28), and (81) through (84). The stepwise sequence was presented to demonstrate the characteristics common to this and Brown's organoborane chemistry. The strongly caustic reaction medium renders oxidizing agents other than water unnecessary in this example. The hydrogen produced in this equilibrium set is consumed in other reactions in the stack that require hydrogen as a reactant. Carbon in the reactant acetate is in the elemental oxidation state. The products consist of reduced species, hydrogen, methoxide, and *tert*-butoxide, along with oxidized carbonate. This reaction is a simple modification of the typical redox—disproportionation to carboxylates.

Hydrolysis can occur before the tertiary carbonolate is formed. The reactant of equation (84) will hydrolyze as in the following equation:

$$[(CH_3O)(OH)\overset{O}{\overset{|}{Al}}-\overset{O}{C}(CH_3)_2]^- + H_2O =$$
$$[(CH_3O)(OH)_2AlOCH(CH_3)_2]^- \quad (86)$$

Neutralization of the products of equation (86) produces methanol and isopropanol or 1-methoxy-2-propanol, thus accounting for other members of the observed product roster.

Organoborane reactions with carbon monoxide apparently yield compounds possessing boron to carbon bonds. All of the reactions we have described in the present discussion yield the oxidized analogous alkoxides. The presence of 2-butene and 1-methoxy-2-butene among the products of this aluminum oxide modified disproportionation can only be understood if, under certain circumstances, aluminum—carbon bonds can be formed even in these highly caustic conditions. Fortunately, the products are sufficiently explicit that the available reactants delimit the conditions under which the aluminum–carbon bond must be formed.

Metallocyclic diolates and the formation of these from poly-ketene anions under conditions of abnormal water addition have already been described in detail. Reaction of the aluminum 2-butene-1,4-diolate with methylide could produce the required compounds.

$$[(H_3CO)_2\overline{AlOCH_2CH{=}CHCH_2O}]^- \;+\; CH_3^- \;=\;$$
$$CH_3O^- \;+\; [(H_3CO)_2\overline{AlCH_2CH{=}CHCH_2O}]^- \quad (87)$$

Acid hydrolysis of the products of equation (87) would yield methanol and 1-methoxy-2-butene.

$$[(H_3CO)_2\overline{AlOCH_2CH{=}CHCH_2O}]^- \;+\; 2CH_3^- \;=\;$$
$$2CH_3O^- \;+\; [(H_3CO)_2\overline{AlCH_2CH{=}CHCH_2}]^- \quad (88)$$

The hydrolysis of the products of equation (88) would yield methanol and 2-butene.

The unidentified butanol listed among the products of Table 4.11 can be *sec*-butanol or 2-methylpropanol. The presence of 2-methoxybutane in the product roster supports the assignment of the *sec*-butanol for this butanol. The absence of any *n*-butanol or its ethers possibly signifies that the *sec*-butanol is a product of some rearrangement such as that previously described.

The product rosters of the many reactions studied have contributed to this equilibrium model. The disproportionation of carbon is a complex system of interconnected reactions like those discussed here. As is true of most reactions in aqueous media, the reactions are fast. At very high activities of base the reaction solvent, as previously pointed out, is essentially molten caustic. While this is not an *aqueous* solvent, equilibrium is apparently reached rapidly as a consequence of the intrinsically high ionic strength.

VAPOR PHASE COMPOSITION

The composition of the autogenous vapor phase in equilibrium with the entire reaction stack is of interest. The difficulties in getting quantitative data describing this composition have already been mentioned in passing. The pressure behavior of the many different systems studied permits informed estimates of this composition.

For the typical disproportionation of graphite, or cellulose, in molten potassium hydroxide, with moderate water activity, the vapor phase at $360°C$ is mainly water mixed with less than 5 mol% of carbon monoxide and hydrogen combined. When carbonate is used as the base, the observed autogenous pressures are nearly double that of the potassium hydroxide reaction. The additional pressure is mainly due to carbon dioxide in the vapor phase. In these reactions there are also some lower hydrocarbons contributing to the pressure. Since the pressures of most of these reactions fall to vacuum upon cooling to room temperature, these hydrocarbons are in equilibrium with the reaction stack. Olefins are the most likely hydrocarbons to behave in this fashion. The modified reaction in potassium hydroxide, in which manganese dioxide, boric oxide, and aluminum oxide are employed as adducts, also displays higher pressures. In all of these reactions acetate is absent from the roster of products. Acetate results from the simple hydration of carbon, with no redox–disproportionation required. When acetate is consumed in the modified reactions, additional carbonate is formed as in equation (85). The carbonate–bicarbonate equilibrium is activated, and additional carbon dioxide consequently appears in the vapor phase, increasing the observed autogenous pressure.

One of the ways to assess the role of the various equilibria involving a gaseous reactant is to perturb the pressure of that reactant and determine the effect of that perturbation upon the product roster. Experiments involving both hydrogen and carbon monoxide have shed some light on early reactions in the stack.

The reaction of carbon monoxide with either sodium or potassium hydroxide under autogenous pressure can be carried out in such a fashion that carboxylates from acetate to caproate are the main products. After the high-temperature reaction is complete and the product mixture cooled to room temperature, a very small pressure of hydrogen is found in the gas phase. In the reactant mixture a similarly small amount of hydride is measured.

It is interesting that no formate or oxalate is found among the products of most redox–disproportionations of carbon. When carbon monoxide is one of the carbonaceous reactants, however, formate is detected, but still no oxalate.

In addition to the hydrogen and hydride described above, carbon dioxide is also found in the gas phase. This requires the presence of appreciable activity of bicarbonate ion in the solution. In alkaline solutions the following equilibrium adjusts the activities of carbonate and bicarbonate in the presence of carbon dioxide:

$$CO_3^{2-} + CO_2 + H_2O = 2HCO_3^- \tag{89}$$

The small amount of hydrogen found in the gas phase after the reaction is complete is an artefact of the microreactor system. The reason for this systematic problem was discussed earlier in the chapter. It is possible, however, that gaseous hydrogen represents an equilibrium pressure of hydrogen at $360°C$. This assumption is consistent with the observation that addition of hydrogen to the carbon monoxide reactant prevents any measurable reaction. These facts suggest several interconnected equilibria. The following equations illustrate one set that would behave in this interesting manner:

$$CO + OH^- = HCOO^- \tag{90}$$

$$CO + OH^- + H_2O = H_2 + HCO_3^- \tag{91}$$

$$H_2 + OH^- = H_2O + H^- \tag{92}$$

$$H^- + HCO_3^- = HCOO^- + OH^- \tag{93}$$

$$2HCOO^- = {}^-OOC-COO^- + H_2 \tag{94}$$

The dramatic effect that hydrogen has in this case requires it to be a participant in an important early reaction. The most likely candidate for this is the formate-oxalate equilibrium shown in equation (94). Experiments suggest that the partial pressure required for this suppression is quite low. At $250°C$ under a total pressure of 250 psig a mixture of carbon monoxide and hydrogen with the mole fraction of hydrogen less than 0.02 will suffice. This corresponds to an equilibrium pressure for equations (91) and (94) of approximately 250 torr. This constitutes an upper limit for the amount of hydrogen present in the vapor phase at $360°C$. The estimate that carbon monoxide and hydrogen *combined* comprise no more than 5 mol% of the autogenous equilibrium pressure is seen to be quite conservative.

The reaction of carbon monoxide with molten caustic is known to produce both formate and oxalate in reactions at or near atmospheric pressure.[5,23-27] Below $450°C$ the reaction product is mainly formate; above that temperature, oxalate is formed. This fact would be consistent with an equilibrium hydrogen pressure of one atmosphere at $450°C$ for equation (94). Furthermore, if molten potassium hydroxide is used, formate formation is apparently favored. The use of sodium hydroxide, or mixtures of sodium and potassium hydroxide apparently favors oxalate.

The difference in behavior of potassium and sodium hydroxide is due to a frequently overlooked fact. Pure solid sodium hydroxide contains only 2% water by weight. Pure solid potassium hydroxide is 15% water by weight. The water activity can be adjusted by using one or the other or by using different mixtures of both. Equations (91) and (92) tie the activities of water, hydroxide ion, and hydrogen together in such a fashion that the effect of increasing the water activity is to increase the activity of hydrogen. Thus water has the effect of shifting the equilibrium described in equation (94) to favor formate.

When carbon monoxide is added to molten potassium hydroxide at $350°C$ in a closed system, only formate and a trace of oxalate are produced. All of the carbon monoxide consumed is accounted for by these two products. It is clear from the set of equations (90) through (94) why this should be the case. There is no equilibrium *consuming* hydrogen that would connect this set of equilibria with those involving any species more complex than oxalate. The equilibrium system as a consequence can proceed no further than oxalate. The amount of oxalate at equilibrium in a closed microreactor is proportional to the pressure of carbon monoxide in equilibrium with the potassium hydroxide melt.

This condition is immediately changed by the addition of small amounts of cellulose to the alkaline reactant. This produces intermediate carbon species, possibly ketene anions, in solution. Addition of a small amount of graphite produces the same result. Graphite is in equilibrium with such species, and the resulting stack of equations is identical to that for graphite as the carbonaceous reactant. Once established, this stack can consume more carbon monoxide and produce more complex carboxylates.

The presence of graphite after the establishment of the equilibrium in the cellulose reaction summarized in Table 4.12 signifies that ketene anion is a viable reactant. The equilibrium in the following equation could fix the activity of the formate, maintaining oxalate at levels too small to detect among the products:

Table 4.12 Complete Carbon Balance for Cellulose in Molten Potassium Hydroxide[a]

Reagents: (g/mmol)[b]
85% KOH = 2.48/37.6; H_2O = 1.4 (est.)/76; $C_6(H_2O)_5$ = 0.15/5.65
CO (consumed) = 1.49 mmol

Products	Yield[c]	
	mol%	wt%
HCOOH	12.6	9.2
CH_3COOH	59.4	56.3
C_2H_5COOH	21.6	25.2
C_3H_7COOH	5.2	7.2
C_4H_9COOH	1.3	2.1
$C_5H_{11}COOH$	trace	trace
C_6H_5OH	trace	trace
$CH_3C_6H_4OH$	trace	trace

[a] High Water Activity (under 200 psig carbon monoxide).
[b] CO_2 accounts for 0.49 mmol of initial carbon, reduced products for 4.24 mmol.
[c] Conversion = 79.7%; Material Balance = 100%.

$$HC{\equiv}CO^- + HCO_2^- + H_2O = H_2C{=}CHCO_2^- + OH^- \qquad (95)$$

The reduction of the acrylate to propionate by hydrogen would be the likely result in these solutions. All of the equilibria maintaining ketene(polyketene) anions furnish analogous reactants. These formate addition reactions all would produce odd-numbered carboxylates. It is perhaps significant that, for the experiments utilizing carbon monoxide in combination with some other carbonaceous reactant, the product rosters appear to be richer in odd-numbered carboxylates. These results are most easily rationalized by invoking formate additions like that shown in equation (95).

CONCLUSIONS

The utility of the ketene(polyketene) reaction model in accounting for the observed chemistry is quite persuasive. The positions of the carbon solubility equilibria, establishing the equilibrium activities of these various reactants, as in equations (34), (45), and (48), must lie well to the left. Apparently the low solubility of graphite in molten potassium hydroxide is the rate-limiting equilibrium in this reactant system. Otherwise, it is difficult to rationalize the low rate of conversion of graphite in 8 hr (14%) compared to the high conversions of cellulose to products.

The constant autogenous pressure observed in these reactions reflects the constancy of all reactant descendants present in the liquid phase. The same reactant descendants apparently are obtained whether graphite, carbohydrates, or carbon monoxide is the carbonaceous reactant. With minor variations, all carbonaceous reactants establish similar reaction stacks once the equilibrium reaches its clamped position.

Where possible, individual equilibria must be isolated from the stack and examined independently. Information gained from such experiments can only refine the admittedly crude mechanistic model presented here.

The practical result of this experimental work is not insignificant. It does provide an upper limit beyond which the price of crude petroleum cannot go without competition from this chemistry. The experimental work described and discussed in Chapters V and VI could lower this upper limit significantly.

None the less, as with most research directed toward practical goals, the most important result of this work is the new chemistry learned. The novel aluminum alkyl chemistry used to account for some of the product roster in Table 4.11 is particularly interesting. In the equilibria described in equations (81) through (88), it is apparent that no gases are involved as reactants in any of these. The vessels required to contain the autogenous pressure of these reactions are not required for these particular equilibria. Equation (88) will proceed to the right more completely at lower pressures, since hydrogen is a product of this equilibrium as shown.

It is possible to conceive of benchtop experiments utilizing these equilibria. Acetate ion and aluminate ions are required as reactants under conditions of low water activity. The water activity must be as low as that of a molten potassium hydroxide solution. While this requirement is rigorous, it is possible to combine these reactants in solvents that accomplish this. Experiments to produce useful amounts of tertiary carbinols are in progress and will be discussed later in a different forum.[28]

REFERENCES

1. R. Hare, *Am. J. Sci.*, **37**(1), 268–269, (1839).
2. H. Feuchter, *Chem.-Ztg.*, **38**, 273–274 (1914).
3. H. S. Fry, E. L. Schulze and H. Whitkamp, *J. Am. Chem. Soc.*, **46**, 2268–2275 (1924).
4. J. Liebig and F. Wöhler, *Ann.*, **3**, 249–282 (1832).
5. J. B. A. Dumas and J. S. Stas, *Ann. Chim. Phys.*, **35**, 129–174 (1840).
6. M. P. Crimmins and G. Urry, unpublished results.
7. C. Weizmann and S. F. Garrard, *J.C.S.*, **117**, 332–338 (1920).
8. H. Okamoto, M. Katsumoto, and T. Kudo, *Bull. Chem. Soc. Jpn.*, **56**, 925–926 (1983).
9. D. Davidson and M. T. Bogert, *J. Am. Chem. Soc.*, **57**, 905 (1935).
10. F. R. Curadau, *Ann. Chim. (Paris)*, **66**, 97–103 (1808).

11. J. L. Gay-Lussac and J. L. Thenard, *Recherches Physico-Chimiques*, Paris, 1, 101 (1811).
12. J. P. Kakareka, R. Y. Weng, and G. Urry, in preparation.
13. H. Meerwein and R. Schmidt, *Ann.*, **444**, 221–238 (1925). See also: W. Ponndorf, *Z. Angew. Chem.*, **39**, 138–143 (1926) and A. Verley, *Bull. Soc. Chim. Fr.*, **37**, 537–542 (1925).
14. A. L. Wilds, in *Organic Reactions*. Vol 2, R. Adams et al., Eds., John Wiley and Sons, New York, 1944, Chap 5.
15. R. V. Oppenauer, *Recl. Trav. Chim. Pays-Bas*, **56**, 137 (1937).
16. V. E. Tishchenko, *J. Russ. Phys.-Chem. Soc.*, **38**, 355, 482 (1906).
17. J. C. Kleingeld and N. M. M. Nibberung, *Int. J. Mass Spectrom. Ion Phys.*, **49**, 311–318 (1983).
18. J. F. Paulson and M. J. Henchman, *Bull. Am. Phys. Soc.*, **27**, 108 (1982).
19. W. M. Vogel, K. J. Routsis, V. J. Kehrer, D. A. Landsman, and J. G. Tschinkel, *J. Chem. Eng. Data*, **12**, 465–472 (1967).
20. F. Varrentrap, *Ann. der Chem. Pharm.*, **35**, 196–215 (1840).
21. R. Hyatsu, R. L. McBeth, R. G. Scott, R. E. Botto, and R. E. Winans, *Org. Geochem.*, **6**, 463–471 (1984).
22. H. C. Brown, *Organic Syntheses via Boranes*, John Wiley & Sons, New York, (1975), Chap. 7.
23. P. E. M. Berthelot, *Ann. Chim. Phys.*, **46**, 477–487 (1856).
24. V. Merz and J. Tiberica; *Ber. Dtsch. Chem. Ges.*, **13**, 23–33 (1880).
25. V. Merz and W. Weith, *Ber.*, Dtsch. Chem. Ges. **15**, 1507–1512 (1882).
26. E. H. Leslie and C. D. Carpenter, *Chem. Metall. Eng.*, **22**, 1195–1197 (1920).
27. M. C. Boswell and J. C. Dickson, *J. Am. Chem. Soc.*, **40**, 1779–1796 (1918).
28. J. P. Kakareka and G. Urry, work in progress.

Chapter V

Nitrogen Fixation by Carbon

HISTORICAL REVIEW OF NITROGEN FIXATION

In the recent past fixation of nitrogen became a highly desirable goal of many chemists interested in novel applications for homogeneous catalysts.[1] In this energetic effort it apparently was a rule of the game that such chemistry should, if possible, emulate biochemical fixation. This was, in practice, the *raison d'etre* for the initial interest in such processes. Persuasive economic arguments abounded in the pursuit of this goal.

This effort was doomed to disappointment as a competitive process since the economics of the Haber process[2] were so difficult to overcome. The direct synthesis of ammonia under conditions of heterogeneous catalysis is relatively economical of energy and for many years enjoyed the advantage of utilizing reactants that were literally as free as air. During the long homogeneous catalysis research effort the goal seemed to become more general until chemists were conditioned to view this as both challenging and elusive. As in many other cases the chemistry learned in this search is much more important than the lack of any practical result indicates. Knowledge of bonding principles

149

and detailed reaction mechanisms for many transition metal coordination complexes increased considerably in the course of this well-staffed, well-funded effort.[1]

The repetition of history seems to arise naturally from the human condition even as times change. The abrupt changes arising from actions of the petroleum cartel in the middle 1970s produced yet a different situation. Hydrogen, no longer a worthless by-product of petroleum refining, was used to the maximum to increase yields of more valuable petroleum fractions. Its economic worth, instead of being based mainly upon a value in use as a component of ammonia, assumed a hard and measurable competitive value in gasoline. The effect of this change upon the costs of agricultural production are a recent painful memory.

The fact that chemical nitrogen fixation traditionally utilized carbon exclusively is apparently lost in the dim past of chemistry.[3-5] The cyanamide process[6] for ammonia and the Bucher process[7] for the manufacture of sodium cyanide and ammonia have, for good reasons, been viewed as excellent and competitive processes for fixation of nitrogen.[8] Of these the most interesting is the Bucher process, which involves heating a mixture of anhydrous sodium carbonate and coke at temperatures near $1000°C$ in a stream of air or nitrogen.

The major reasons why the Haber process dominated in nitrogen fixation, aside from the previously discussed availability of cheap hydrogen, were process difficulties encountered in large-scale applications of Bucher's method. The chemistry of the Bucher process is summarized in the following equation:

$$Na_2CO_3 + 4C + N_2 = 2NaCN + 3CO \qquad (1)$$

Under the conditions of the Bucher process, temperatures in excess of $850°C$ are required to initiate this reaction. After an accumulation of cyanide product the reaction mass becomes a viscous liquid. Sodium carbonate and sodium cyanide form a $Na_2CO_3-4NaCN$ eutectic which melts at $460°C$.[9] Heat transfer required in this overall endothermic reaction is inefficient under these conditions and as a consequence higher reaction temperatures are needed to complete the cyanation. The heating of the reaction mass is mainly by internal burning of reactant carbon, since the nitrogen source is air. At the higher temperatures silica interferes with the reaction. This is possibly a consequence of the stabilization of the graphite phase by edge silicon carbide, as discussed in Chapter II. The handling of a viscous liquid mass in a continuous process presents formidable, almost insurmountable, difficulties. Initiation of the reaction apparently requires heating of the sodium carbonate coke mass to temperatures in the region of the melting point of the carbonate.

In Chapter II the nature of the graphite phase was discussed. The passivating ketonic edges also were shown to require temperatures near $800°C$ in order to activate the graphite. In Chapter IV the activation of graphite at temperatures below $400°C$ was shown to be rapid in molten hydrates of various hydroxides as well as similar melts of sodium carbonate. From this information it is deduced that in the Bucher process the high temperature required for initiation depends mainly upon the existence of a molten phase and the activation of the graphitic phase would be rapid under such conditions. It was reasoning such as this which stimulated the interest in the high-pressure reactions discussed in Chapter IV.

NITROGEN FIXATION AND THE EQUILIBRIUM DISPROPORTIONATION OF GRAPHITE

In the preceding chapter we described several reactions in which the redox–disproportionation of graphite to carbonate and a mixture of carboxylates was facilitated in molten potassium hydroxide. In these reactions water was added in various amounts. In Table 5.1 several reactions, in which the graphite, potassium hydroxide, and water were fixed at comparable levels, were carried out under different pressures of nitrogen.

Table 5.1 Reactions of Nitrogen with Graphite in Molten Potassium Hydroxide[a]

Reagents: (g/mmol)

85% KOH = 2.48/37.6; H_2O = ~1.5/~83; $C_{(gr)}$ = 0.52/43.3

N_2 Pressure	N_2 Consumed	Products	
		NH_3	Other
1000 psig	6.8 mmol	1.5 mmol	pyridine, glycine, glutamic acid and other amino acids
500 psig	5.5 mmol	0.5 mmol	pyridine, glycine, glutamic acid and other amino acids
200 psig	2.6 mmol	0.2 mmol	pyridine, quantities insufficient for amino acid analysis
100 psig	1.3 mmol	0.1mmol	pyridine (trace) no further analysis
50 psig	0.7 mmol	0.1mmol	none detected

[a] Moderate Water Activity (under various pressures of N_2)

The nitrogen consumed in these reactions was determined by the methods described in Chapter IV. The ammonia product was determined in the gas phase after the equilibrium reaction was complete. A portion of the alkaline reaction product was analyzed by gas chromatogram mass spectrometry. Pyridine was identified in this analysis. The remainder of the product was acidified and subjected to thin-layer chromatrography, as well as analysis in an amino acid analyzer. The high salt content of the acidified product made the amino acid analysis difficult. Quantitative information was not obtained for any of the products listed in the column labeled "Other".

The quantitative data obtained from this series of experiments clearly characterize the reactions that consume nitrogen as being at equilibrium with the liquid reactants. The pressure dependence of the amount of nitrogen consumed as well as the pressure dependence of the ammonia produced verifies this conclusion.

The chemistry of this reaction needs to be interpreted in the context of the detailed reaction mechanisms suggested in Chapter IV. That series of reactions was most readily interpreted as arising naturally from known ketene(polyketene) anion chemistry. The character of the reaction medium for the examples summarized in Table 5.1 is identical with all of the reactions of graphite effected in molten potassium hydroxide. The changes in the product roster that result from added gaseous nitrogen require the mechanism to include reactions of that molecule with some components of the reaction medium. The products of these reactions should lead sensibly to the roster of products observed.

In describing the nature and activation of graphite in Chapter II, the reaction of active graphite with elemental nitrogen at temperatures as low as $400°C$ was demonstrated. Speculation concerning that reaction invoked the availability of biradical (or polyradical) states in which the stereochemical requirements of the antibonding orbitals of the nitrogen molecule were nicely matched at $400°C$. The polyketene anions present in these reaction media enjoy the same fit. The following equation illustrates such a reaction with the four carbon ketene anion:

$$[O^\nearrow C \searrow C ^\nearrow C \searrow CH]^- \; + \; N_2 \; = \; [O^\nearrow C \searrow C ^\nearrow C \searrow CH]^- \qquad (2)$$
$$\qquad\qquad\qquad\qquad\qquad\qquad\qquad N=N$$

The reactant polyketene anion of equation (2) is presented in the bent form, polyradical state that is most likely under the conditions of this chemistry. The ground state of the parent molecule, butatrienone, is estimated in theoretical calculations to be bent.[10] The product of equation (2) would be stabilized by hydrogenation or hydration as follows:

$$[O^\nearrow C \searrow C ^\nearrow C \searrow CH]^- \; + \; H_2O \; = \; [O^\nearrow C \searrow C ^\nearrow C \searrow CH]^- \qquad (3)$$
$$\quad N=N \qquad\qquad\qquad\qquad\qquad\quad N=N$$

Rearrangement of the 1,2-diazine product of equation (3) to an imidazole is required in order to account for the amino acids formed during the hydrolysis of the alkaline product mixtures summarized in Table 5.1. Many rearrangements of diazines to more stable isomers are known. The benzidine rearrangement[11] and the thermal rearrangement of pyridazine to pyrimidine or pyrazine,[12] illustrate this type of rearrangement. Thermal and photochemical rearrangements of the diazines currently are believed to have similar mechanisms.[13-15] The rearrangements typically are to more stable diazines.

The thermal rearrangement of alkyl isocyanides to nitriles[16] may be related to the rearrangements of the diazines. The heat of isomerization is very low for this reaction, approximately 21 kcal per mole. There is much disagreement about the mechanism for this reaction, but the net process can best be described as an end-to-end rotation of the CN group. This end-to-end rotation is analogous to the currently accepted mechanism for the benzidine rearrangement.[11] It is reasonable to expect chemical moieties that approximate sphericity to undergo such rotations at an appropriate temperature. The rotation can be described as a sequence of bond breaking and bond making, but it is not certain that the molecules themselves are cognizant of such intellectual niceties. Thus, a 1,2-diazine need merely to undergo one such rotation to convert the 1,2-diazine product of equation (3) to an imidazole:

$$[O^{-C}\diagdown_C{}^{-C}\diagdown_C{}^{-C}\diagdown CH]^- = [O^{-C}\diagdown_C{}^{-C}\diagdown_C{}^{-C}\diagdown N]^- \qquad (4)$$

Oxidation of the substituent on the imidazole product of equation (4), followed by decarboxylation, would produce the lactam required for hydrolysis to glycine. Reduction to a methyl substituent would furnish the precursor for alanine. The glutamic acid and the pyridine listed in the product roster of Table 5.1 require longer polyketene anions, such as the six-carbon anion shown in equation (5). There are three different likely products of nitrogen addition in this example.

$$3[O{-}C{\approx}C{-}C{\approx}C{-}C{\approx}CH]^- + 3N_2 = [O{-}C{\approx}C{-}C{\approx}C{-}C{\approx}CH]^- + \\ \overset{|}{N}{=}\overset{|}{N}$$

$$[O{-}C{\approx}C{-}C{\approx}C{-}C{\approx}CH]^- + [O{-}C{\approx}C{-}C{\approx}C{-}C{\approx}CH]^- \quad (5)$$

Hydrogenation and hydration of the diazine analogous to the product of equation (2) could proceed according to equation (6).

$$[O{-}C{\approx}C{-}C{\approx}C{-}C{\approx}CH]^- + H_2 + H_2O = [O{-}C{\approx}C{-}C{\approx}C{-}C{\approx}CH]^- \quad (6)$$

The other two products of equation (5) would hydrogenate and hydrolyze in a similar way to produce the intermediates shown here:

$$[O{-}C{\approx}C{-}C{\approx}C{-}C{\approx}CH]^- + H_2 + H_2O = \\ [O{-}C{\approx}C{-}C{\approx}C{-}C{\approx}CH]^- \quad (7)$$

$$[O{-}C{\approx}C{-}C{\approx}C{-}C{\approx}CH]^- + H_2 + H_2O = \\ [O{-}C{\approx}C{-}C{\approx}C{-}C{\approx}CH]^- \quad (8)$$

The products of these two reactions would necessarily stabilize in a different fashion than the products of equations (3) and (6) since an imidazole, as produced in equation (4), is not likely from a simple rearrangement. More stable intermediates

arise from rearrangements which result in one of the nitrogen
atoms occupying a terminal position.

$$[O^{\diagup C}\diagdown_C{}^{\diagup C}\diagdown_C{}^{\diagup C}\diagdown_{CH}]^- = [O^{\diagup C}\diagdown_C{}^{\diagup C}\diagdown_C{}^{\diagup C}\diagdown_{NH}]^- \qquad (9)$$

$$[O^{\diagup C}\diagdown_C{}^{\diagup C}\diagdown_C{}^{\diagup C}\diagdown_{CH}]^- = [O^{\diagup C}\diagdown_C{}^{\diagup C}\diagdown_C{}^{\diagup C}\diagdown_{NH}]^- \qquad (10)$$

Under the conditions of these high-temperature disproportionations in molten potassium hydroxide the products of both
equations (9) and (10) would rapidly be converted to pyridine
and ammonia.

This series of reactions, equations (2) through (10) reasonably
accounts for the products listed in Table 5.1. They are simple
modifications of the reactions carried out under autogenous
pressure with nitrogen absent.   There is no novel chemistry
required, and the product roster is made rational. The amounts
of each produced would depend upon such factors as relative
solubilities of solid products, relative volatilities of condensible
products, and the positions of the equilibria for each of the
involved reactions in the clamped stack of reactions.

Unfortunately, the nature of the various caustic melts that
serve as the medium for those interesting reactions also requires
the use of pressure reaction vessels of expensive metal alloys.
Thus, while the chemistry is simple and the engineering relatively
straightforward, a production plant utilizing such chemistry would
involve an unusually high capital burden.   An estimate based
upon crude oil costs as the competitive source of liquid hydrocar-

bon places the breakeven point for this process at $40 (1985 dollars) a barrel crude oil price.

The basic requirements of any useful chemical system for this equilibrium disproportionation chemistry are simple. The reactions must be effected in a medium of high ionic strength and relatively high ionic mobility. The reaction phase must be at equilibrium with the gas phase during the course of the reaction. Finally, the equilibrium conditions must be adjustable in order to effect the selectivity of product inherent to this equilibrium system.

OPEN STRUCTURE OF CRYSTALLINE SODIUM CARBONATE

The sesquicarbonate of sodium, $Na_3(CO_3)(HCO_3)$, known as *natron*, one of the oldest chemicals of commerce[17] and mined in practically pure form all over the world, is an interesting material. Part of the interesting behavior of this material arises from the unusual crystal structure of sodium carbonate.

Instinctively, one assumes that the crystal structure for such a classically important chemical would have been long since cleanly established. Perusal of the literature produces the surprising fact that great uncertainty still exists.[18-22]

After some thought, the nature of the sodium carbonate ion pair is seen to be the natural consequence of the fact that sodium ion coordination is octahedral and that of the dinegative carbonate ion, trigonal planar. The requirements of charge neutrality and coordinative saturation for any crystal allows for several different orientations of carbonate ions. Two of the trigonal sites of the carbonate can be coordinated to a single sodium ion, two sites can bridge between two sodium ions in the same layer, or two of the trigonal sites can bridge between two sodium atoms in different layers. The principle of charge neutrality rules out the site where all three ligancies of the carbonate are coordinated to a single sodium ion. Such coordination would probably create an f-center in the crystal. There are six possible different permutations of the three carbonate ligancies in this crystal. In the sodium carbonate crystal these types of carbonate sites are apparently disordered and nearly random.

Aside from creating difficulties for the crystallographer, the nature of the sodium carbonate crystal produces some novel properties for the carbonate and the sesquicarbonate. A highly compressed pellet of the sesquicarbonate can be dehydrated readily at 300°C to pure sodium carbonate with no change in the physical dimensions of the pellet. The sodium carbonate resulting from this dehydration is considerably less dense than the original sesquicarbonate yet displays the characteristic x-ray patterns for crystalline carbonate.

EQUILIBRIUM REACTIONS IN SOLID
SODIUM CARBONATE

The open structure of sodium carbonate results in useful chemical behavior in the solid state. It is possible to convert pure solid sodium carbonate into pure solid sodium cyanide.[23] Moist carbon dioxide can reverse this reaction.[24,25] The reversibility of this reaction implies an equilibrium process with equilibrium between the entire solid phase and the gaseous phase. A more thorough examination of the pressure-composition dependence of this reaction supports such a conclusion as well.[26]

The reaction described must be effected at temperatures below $450°C$ since there exists a 47% sodium cyanide–53% sodium carbonate (by weight) eutectic which melts at $456°C$.[6] For conversions of carbonate to cyanide this viscous liquid state must be avoided since mass transport and heat conduction become inefficient.

Sodium carbonate enjoys another interesting virtue in the context of base-induced disproportionations. The following temperature dependent equilibrium should allow alteration of the virtual basicity in the crystalline state:

$$Na_2CO_3 \ = \ Na_2O \ + \ CO_2 \tag{11}$$

Below $400°C$ carbon monoxide gas equilibrates with sodium carbonate in its open crystalline state. The reaction is the classic carbonate–oxalate equilibrium:

$$Na_2CO_3 \ + \ CO \ = \ Na_2C_2O_4 \tag{12}$$

Heating pure sodium oxalate, under an atmosphere of argon, produces little carbon monoxide. At approximately $420°C$ it melts to a water clear liquid. This liquid rapidly turns black. The following equation describes this pyrolytic decomposition of sodium oxalate:

$$2Na_2C_2O_4 \ = \ C_{(gr)} \ + \ 2Na_2CO_3 \ + \ CO_2 \tag{13}$$

This black liquid continues to decompose, with carbon dioxide evolution continuing for some time, until solidification to a black solid. It may be that oxalate also forms a eutectic with sodium carbonate.

The question arises whether the graphite produced under these conditions is in an active or passivated form. In Chapter II it was determined that active graphite was capable of reacting with molecular nitrogen at temperatures of $400°-450°C$. When the molten pyrolysis of oxalate is carried out under an atmosphere of nitrogen, the black solid mass remaining after the pyrolysis contains cyanide. It appears that the graphite deposited in a matrix of sodium carbonate is in an active form and that it is capable of undergoing reactions typical of active graphite.

APPARATUS AND PROCEDURES

The pyrolytic decomposition of sodium oxalate is an inefficient way to produce sodium carbonate with an intimate admixture of active graphite. Furthermore, the difficulties of establishing equilibrium between a viscous liquid and a gas have already been discussed. A method of producing a dispersion of active graphite in a lattice of open structured solid sodium carbonate is clearly to be preferred. Such a dispersion is readily prepared by mixing stoichiometric quantities of sodium bicarbonate and sodium oxalate. Pressing this mixture into a pellet followed by the thermal decomposition of the pellet at $300°C$ produces a mixed crystalline solid containing the carbonate and oxalate. Pyrolysis of this pellet at $450°C$ produces the desired dispersion of active graphite in an open structured sodium carbonate crystal.

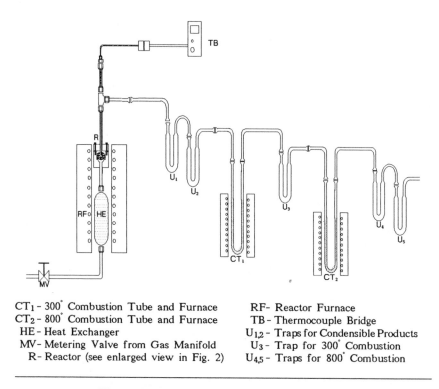

CT_1 - 300° Combustion Tube and Furnace   RF- Reactor Furnace
CT_2 - 800° Combustion Tube and Furnace   TB - Thermocouple Bridge
HE - Heat Exchanger $U_{1,2}$ - Traps for Condensible Products
MV - Metering Valve from Gas Manifold U_3 - Trap for 300° Combustion
 R - Reactor (see enlarged view in Fig. 2) $U_{4,5}$ - Traps for 800° Combustion

Figure 5.1 Gas Flow Reactor and Analysis Train

These procedures require special apparatus and the general form of this is presented in Fig. 5.1. The pellet of reactants is prepared by mixing the desired components in a mortar. The intimate mixture thus produced is placed in a die and compressed by means of a hydraulic press under a pressure of 10,000 psi. The resulting pellet is cylindrical with spherically convex ends. Its radius is 13 mm with an overall height of approximately 5-8 mm. The weight usually falls in the range of 1-2 g. The pellet is then drilled such that the hole produced is centrally located, with the bottom of the hole just below the center of the pellet.

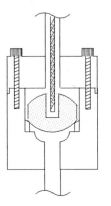

Figure 5.2 Reactor Detail

The diameter of the hole is chosen to provide a close fit to the temperature probe illustrated in the figure. The temperatures determined during the experiment are those at the center of the carbonate pellet.

The prepared pellet is mounted in the reaction chamber, as shown in Fig. 5.2. It is heated by means of a stream of gas which passes through a heat exchanger filled with a fluidized bed of silicon carbide sand. This heat exchanger is surrounded by a muffle furnace. The temperature of the muffle furnace is controlled by a variable transformer. This temperature is adjusted to produce the desired temperature at the center of the pellet. Typical flow rates of 100-200 cc per minute are sufficient for most experiments.

Different gases can be used in this heating procedure. For example, during the decomposition of the bicarbonate at $300^{\circ}C$ and the subsequent pyrolysis reaction at $450^{\circ}C$, argon is used. The active graphite dispersion in solid open-structure sodium carbonate can then be treated with various reactant gases, or mixtures of gases, at a pellet temperature of choice.

While this temperature is not necessarily the temperature of the gases surrounding the pellet, it is a more significant and reproducible temperature. Different gases at the same temperature could produce quite different pellet temperatures because of differences in heat content and conduction. The interior temperature of the pellet is symptomatic of the average temperature of the entire pellet. The actual reaction temperatures, however, could be somewhat higher.

All of these novel methods are simple modifications of the Bucher process. Improved knowledge of the active forms of graphite and their reactions, as discussed in Chapter II, permitted informed modifications. The formation of cyanide in the reaction of nitrogen and active graphite in the temperature range of $400°–450°C$ avoids the viscous melt difficulties of the original process. The open structure of pure sodium carbonate when it is prepared by the decomposition of the bicarbonate or the sesquicarbonate enables the intracrystalline equilibrium chemistry required.

The equilibration of moist carbon dioxide with the solid cyanide–carbonate avoids the expense of mineral acids necessarily borne by the Bucher process. The changes should render this simple chemistry an economical method for the preparation of hydrogen cyanide. This low-temperature fixation of nitrogen also makes it an excellent candidate for the preparation of other derivative nitrogen compounds.

The solid-state equilibration processes also appear to afford excellent means of avoiding pollutants since the simplicity of these processes precludes the need for many noxious processing chemicals like sulfuric acid. In terms of carbon dioxide pollution it does not suffer in comparison with the present method of producing hydrogen cyanide from ammonia and carbon monoxide. The ammonia requires hydrogen from the water-gas reaction and the water gas shift reaction that produces equivalent amounts of carbon dioxide.

GRAPHITE DISPERSIONS IN THE CARBONATE LATTICE

Since the sole role for oxalate in the chemistry described is to produce an intimate dispersion of active graphite in the carbonate lattice, it is easily replaced with any substance capable of decomposition to active graphite. The simplest such substitution is cellulose. Pellets of sodium bicarbonate and crystalline cellulose, prepared as described, decompose at a pellet temperature of $450°C$ to a dispersion of active graphite in sodium carbonate.

Over a century ago a mixed solid, prepared by intimately mixing potassium bicarbonate and sucrose, was decomposed by heating in a stream of air, from which water and oxygen had been removed by means of a chemical absorption train.[27] This mass was converted to potassium cyanide by a long period of heating. The heating was by means of hot nitrogen gas, at a temperature described as white heat. This should correspond to the range of $1000°-1100°C$. At a temperature this high the reaction mass must surely have been liquid. This method was described as suitable for large scale reactions although no information was given for the time required. The chemistry discussed in this chapter is the same as the classical work cited, with novel methods for effecting the reaction at much lower temperatures.

The behavior of pure sodium oxalate upon rapid heating has been described earlier. It does not liquefy until it reaches a temperature above $400°C$ in spite of its reported melting, with decomposition, at $275°-285°C$. It also liquefies without appreciable carbon monoxide evolution. According to the reverse of the equilibrium in equation (12), sodium carbonate is formed *in situ* as oxalate is decomposed:

$$Na_2C_2O_4 \quad = \quad Na_2CO_3 \quad + \quad CO \qquad (15)$$

The concentration of carbonate increases until the mixture becomes a translucent, colorless, fluid liquid. A rapid evolution of carbon dioxide ensues as the liquid turns black and continues as the black liquid solidifies. When the evolution of carbon dioxide is complete, the solid consists of a dispersion of graphite in solid sodium carbonate. The pertinent equilibrium for this step is presented in equation (13).

The basis for suspecting a eutectic of carbonate and oxalate is the fact that a pellet of intimately mixed equimolar quantities of sodium carbonate and oxalate does not melt at $450°C$. The solid pellet held at this temperature for several hours decomposes to a dispersion of carbon in sodium carbonate without changing its physical dimensions. This behavior has greater significance as well.

The formation of a graphite phase as oxalate decomposes in the solid state is an interesting chemical process. During this process apparently no carbon monoxide escapes the crystal lattice. The reduction of oxalate to graphite must be occurring in the oxalate phase. It may be that the oxalate phase is molten and dispersed throughout the expanded sodium carbonate crystal. This is not probable since active carbon in a molten phase fixes nitrogen more rapidly than does this system. The cyanide formation is quite slow in these graphite dispersions, compared with the reactions of graphite in molten potassium hydroxide. It is yet slower than the reactions of nitrogen with cellulose in molten potassium hydroxide. These solutions, prior to reaching equilibrium, possibly contain high concentrations of active carbon species in solution, as discussed previously. Molten sodium oxalate containing active carbon should be rapidly efficient in fixing nitrogen. It should be most like the reactions of cellulose in molten potassium hydroxide under nitrogen.

At present, the best probable conclusion is that the graphite phase grows as an inclusion in the solid phase. Under such conditions, a volatile reducing agent capable of permeating the free space of the crystal would be a serendipitous choice for a mechanism. The decomposition of oxalate to graphite in these pellets is quite rapid at $450°C$. The pellets of dispersed carbon are apparently homogeneous.

It is possible to generate a dispersion of carbon in open structured sodium carbonate by a very simple process. When a pellet of sodium bicarbonate is decomposed to sodium carbonate at a pellet temperature of $300°C$ and then heated at $450°C$, in a stream of pure carbon monoxide, graphite is deposited in the

sodium carbonate. It is apparently uniformly dispersed throughout the crystal. The amount deposited is proportional to the time
of reaction, but the formation of the carbon phase enjoys a
moderate rate. The amount of carbon produced in 4 hr of
treatment at $450°C$ is essentially the same as that produced in the
decomposition of a pellet containing equimolar amounts of
carbonate and oxalate. This experiment demonstrates that carbon
monoxide is an adequate and sufficient reducing agent.

Equation (12) must be the initial step in the formation of this
carbon dispersion. There must be subsequent reactions in which
further carbonyl insertions occur, such as the following:

$$^-O\overset{O}{\overset{\|}{C}}-\overset{O}{\overset{\|}{C}}O^- \ + \ CO \ = \ ^-O\overset{O}{\overset{\|}{C}}\diagdown C\diagup \overset{O}{\overset{\|}{C}}O^- \tag{16}$$

$$^-O\overset{O}{\overset{\|}{C}}-\overset{O}{\overset{\|}{C}}O^- \ + \ 4CO \ = \ ^-OC\diagdown \overset{O}{\overset{\|}{C}}\diagdown C\diagup \overset{O}{\overset{\|}{C}}\diagdown C\diagup \overset{O}{\overset{\|}{C}}O^- \tag{17}$$

In Chapter II the 1,3-ketonic edge of graphite was described
as being reduced readily by gaseous carbon monoxide at temperatures in the range of $400°-600°C$. The 1,3-polyketone dianion
products of equations (16) and (17) should be similarly susceptible
to carbon monoxide reduction:

$$^-OC\diagdown \overset{O}{\overset{\|}{C}}\diagdown C\diagup \overset{O}{\overset{\|}{C}}\diagdown C\diagup \overset{O}{\overset{\|}{C}}O^- \ + \ 4CO \ = \ ^-O\overset{O}{\overset{\|}{C}}-C\equiv C-C\equiv C-\overset{O}{\overset{\|}{C}}O^- \ + \ 4CO_2 \tag{18}$$

Unfortunately, reduction to polyethynyl structures, such as the product of equation (18), is unlikely to be of assistance in forming a graphite lattice. In Chapter I, the likely polymerization of such chemical species to macrocyclic helical structures was cited.[28] Equation (19) produces a more likely candidate for graphite formation in this instance.

$$^-OC \overset{\overset{O}{\|}}{C} \overset{\overset{O}{\|}}{\underset{\overset{\|}{O}}{C}} \overset{\overset{O}{\|}}{\underset{\overset{\|}{O}}{C}} \overset{}{\underset{\overset{\|}{O}}{C}} \overset{\overset{O}{\|}}{C} O^- = OC{=}C{=}CO + {}^-O \overset{\overset{O}{\|}}{C} - \overset{\overset{O}{\|}}{C} O^- + CO_2 \quad (19)$$

The carbon suboxide, C_3O_2, decomposes readily, ultimately to form graphite and carbon dioxide.[29] This reaction questions the wisdom of using these graphite dispersions in attempts at large-scale reactions of gaseous nitrogen. The chemistry described in Chapter II leads to the conclusion that virtually all reactions of active graphite lead to passivation of the graphite phase. The cyanide detected in the examples so far described must arise from the small amount of cyanogen typically produced when active graphite reacts with elemental nitrogen at $450^\circ C$. No more cyanogen can be expected unless some reactant, capable of reactivating the graphite, is added to the system. The best candidate for a gaseous reactant is carbon monoxide. Addition of carbon monoxide to the nitrogen stream in these sodium carbonate reactions should be reflected in increased yields of cyanide.

The other possibilities raised by this mechanistic analysis, however, may mean that it is counterproductive to establish a graphite dispersion in this system. There are implications that, if C_3O_2 is required to form graphite, it is availabe for other reactions. It is a sufficiently reactive molecule to combine with molecular nitrogen in the adsorbed phase.

FIXATION OF NITROGEN BY TRICARBON DIOXIDE

The suboxide, C_3O_2, bears a close similarity, structurally, with butatrienone. Like the trienone, it should have a bent structure with polyradical character. Initially, electron diffraction studies were interpreted as being consistent with a rigidly linear structure.[30] Later application of the same technique suggested that it was a linear molecule that could be bent up to $9°$ at room temperature.[31] Certain anomalies in the infrared spectrum suggest that it is only a quasilinear molecule with a low-frequency bending at the central carbon atom.[32] The bent polyradical structure at elevated temperatures remains probable and is useful in explaining the observed chemistry.

The reaction of tricarbon dioxide with molecular nitrogen is directly analogous to the reactions of ketene anion, proposed as intermediates in the high-pressure reactions with nitrogen summarized in Table 5.1:

$$O^{\cdot C}C^{\cdot C}O \; + \; N_2 \;\; = \;\; O^{\cdot C}C^{\cdot C}O^{\cdot \underset{N=N}{}} \tag{20}$$

Equation (20) describes an equilibrium, like all others in this monograph. In the solid state the reactions available for removing the nitrogen adduct and shifting the equilibrium to the right are limited. One likely reaction is the following:

$$\underset{O^{\diagdown}C^{\diagup}C^{\diagdown}C^{\diagup}O}{\overset{N=N}{}} + CO_3^{-2} = CN^- + CNO^- + 2CO_2 \quad (21)$$

In this reaction system, the cyanate ion will be reduced to cyanide by the reactant gas, carbon monoxide, as follows:

$$CNO^- + CO = CN^- + CO_2 \quad (22)$$

Another likely sequence involves the decomposition of the nitrogen adduct into cyanogen and carbon dioxide:

$$\underset{O^{\diagdown}C^{\diagup}C^{\diagdown}C^{\diagup}O}{\overset{N=N}{}} = NC-CN + CO_2 \quad (23)$$

The cyanogen then could react with carbonate to form cyanide and cyanate, followed by reduction of the cyanate by carbon monoxide. Cyanogen could also react directly with oxalate.

$$NC-CN + C_2O_4^{-2} = 2CN^- + 2CO_2 \quad (24)$$

Sodium carbonate apparently serves as an heterogeneous catalyst for the reactions that involve only gaseous reactants and products. It is a particularly innocuous substance for such an important role.

ATYPICAL PROPERTIES OF THE OPEN
SODIUM CARBONATE LATTICE

Several attributes of these reactions require properties of the expanded lattice sodium carbonate that are not typical of most catalysts. The decomposition of oxalate in a solid pellet, formed by compressing an equimolar mixture of sodium oxalate and sodium bicarbonate, produces no carbon monoxide, even when it is slowly heated to its melting/decomposition temperature and above. No gas is evolved until graphite appears and carbon dioxide produced. During this heating, water and carbon dioxide are expelled from the pellet as the bicarbonate decomposes.

In the thermal decomposition of sodium bicarbonate, two uninegative bicarbonates decompose to form a dinegative carbonate ion, a water molecule, and a carbon dioxide molecule. This isoelectric process is observed when the mixed pellet reaches a temperature required for pure bicarbonate. The decomposition of a dinegative oxalate ion into a dinegative carbonate ion and a carbon monoxide is similarly isoelectric, yet no carbon monoxide can be found in this decomposition. For this to be the case, the equilibrium for equation (12) must lie well to the right *within the pellet* with 1 atm *outside the pellet*.

In the case of bicarbonate decomposition a simple classical rationale is simple. The crystal lattice energies of +1/-2 salts are considerably greater than +1/-1 salts. Sodium carbonate is a moderately disordered crystal, while the bicarbonate is more highly ordered. At low temperatures bicarbonate is favored over the carbonate. Once a temperature is reached where the conversion of bicarbonate to carbonate is feasible, the carbonate is clearly favored thermodynamically. The reverse reaction at the same temperature requires much higher pressures than 1 atm.

In order to put the order-disorder effect into perspective, sodium cyanide, a more highly ordered +1/-1 crystal than sodium bicarbonate, is readily converted to sodium carbonate by moist carbon dioxide.[24,25] This process is clearly driven by crystal-lattice energy effects.

Sodium carbonate and sodium oxalate both are +1/-2 salts, and no similar crystal lattice effects can be invoked. Carbonate and oxalate should be exchangeable with little effect upon the crystal lattice energy over a wide range of temperature, especially in a sodium carbonate crystal with an expanded lattice.

There are more subtle effects than simple ionic charge effects in these reactions occurring within an expanded lattice of sodium carbonate. The pyrolysis of a compressed pellet of sodium bicarbonate into pure *crystalline* sodium carbonate, without change in the physical dimensions of the pellet, must create a lattice with sizable voids. This free space in the crystal should differ from the free space in the crystalline sodium carbonate pellet prepared by pyrolysis of the sesquicarbonate. In both of these instances the voids should be reasonably well distributed throughout the crystal. They also should be reasonably uniform in size.

Adjusting the size of the voids in an expanded sodium carbonate crystal is a simple matter. Pyrolysis of compressed mixtures of sodium bicarbonate and sodium sesquicarbonate of different compositions will give average void sizes between those of the pure substances. Smaller voids result from pyrolysis of mixtures of sesquicarbonate and carbonate. This constitutes an ionic system analogous to the expanded alumina lattices obtained by pyrolyses of different hydrated aluminas.

CLATHRATE OR MOLECULAR SIEVE?

Clathrates form as the host crystallizes in the presence of the captive molecule. There is little evidence at the present time to suggest that the captive molecules equilibrate with the surroundings. If this equilibrium cannot be demonstrated, this class of compounds is quite different from the molecular sieves. From current descriptions of clathrates, it would seem remarkable if such equilibration did not occur. It may be that the clathrates so far characterized consist of host lattices that are so thermally unstable that it is not possible to reach temperatures where equilibration can be demonstrated. Until such clathrates are characterized, this class of compounds will have no utility for the chemistry under discussion.

The nature of the mechanism for molecular sieves and the state of the condensed phase in the voids remains a matter of speculation. In the alumina voids the forces are viewed as similar to van der Waals and London dispersion forces. Ion-dipolar and ion-induced dipolar interactions should be added to these in describing condensed phases in the voids of the expanded sodium carbonate lattice. However such sieves function, the effects should be amplified for sodium carbonate.

In the current discussion a simple, almost simple minded, approach could assess the condensed phase as a dense state similar to that produced by lowered temperature or increased pressure. For example, the molecular sieve capable of reducing the partial pressure of carbon monoxide to 10^{-2} atmospheres has the same effect as cooling the carbon monoxide to $-215°C$. Conversely, it would require a pressure greater than 10^4 atmospheres to produce a carbon monoxide phase of the same density at room temperature.

In certain molecular sieves, cooled to $-196°C$, molecular nitrogen at 1 atm, is adsorbed until the density of the condensed phase is 25% greater than the density of liquid nitrogen at that temperature.[33] The compressibility of liquid nitrogen has not been reported, but this should require pressures in the range of 100–1000 atm.

Either of these estimates, would characterize a molecular sieve as a pressure vessel for the adsorbed molecules.

This notion is not so unrealistic as it first appears. Collisional frequencies between molecules in a gas increase as the square of the density. For a real gas the close approach and the collisional process involve the same forces suggested for the interaction between the adsorbed phase and the material of the molecular sieve. Directly-substituting the collisions between an adsorbed molecule and the walls of a void, therefore, has the same effect as using intermolecular collisions in a real gas. The vapor pressure of the liquid state of the adsorbed molecule should provide a crude measure of the *internal pressure* of the molecular sieve.

An examination of the diffusion processes in molecular sieves indicates that total diffusion is effected by the fastest moving molecules among the sorbate molecules.[34] It is reasonable to describe such adsorption systems as a Maxwell's Demon capable of retaining all but the most energetic of the sorbate molecules —In short, a pressure regulator for the adsorbed phase.

The first reaction to this description of a molecular sieve as a regulated pressure vessel is startling. The process of adsorption is taking gas molecules from a system at low pressure into a region of high pressure, from a gas phase into a condensed phase. The natural query at this point is: What thermodynamic changes provide the energy required for this process? Changes in crystal lattice energy of the molecular sieve can occur to account for all or part of this requirement.

Solid sodium hydroxide can be considered a molecular sieve for water. Water moves from a gaseous phase into the more dense solid sodium hydroxide. The behavior of this water-impregnated sodium hydroxide illustrates this crystal lattice energy effect. Water vapor moves from ambient air into the dense phase of solid potassium hydroxide. The crystal lattice energy can be lowered to the point where the crystal melts. This process is known as *deliquescence*.

If, in place of pure sodium hydroxide, a solid solution of sodium hydroxide and sodium hydride is considered as a molecular sieve, a different source of thermodynamic energy is illustrated. This solid will absorb water from ambient air, expelling hydrogen from the solid, without liquefying, as long as any appreciable amount of hydride is left in the solid. The heat of any chemical reaction occurring in the solid furnishes energy. The reaction of sodium hydride to form sodium hydroxide and elemental hydrogen is strongly exothermic. The expansion of gaseous hydrogen as it moves from the dense phase into the gas phase is an additional source of work for the adsorption process.

Pure crystalline sodium hydride could similarly act as a molecular sieve for water. The solid would gradually be converted to solid sodium hydroxide as hydrogen was generated. At room temperature the reaction of solid crystalline sodium hydride with water proceeds to the point where the surface deliquesces. Considerable amounts of hydride remain in the solid at this point. At room temperature neither sodium hydroxide nor sodium hydride are sufficiently permeable to water vapor to act as molecular sieves. At temperatures in excess of 200°C, however, sodium hydride is converted to sodium hydroxide without any deliquescence. Expansion of the highly ordered lattices of these two substances is required before they possess voids sufficiently large for the processes under discussion to occur.

At elevated temperatures it is also possible to convert solid sodium hydroxide to solid sodium carbonate, merely by passing gaseous carbon dioxide over the solid. This requires temperatures greater than 300°C. Below this temperature, but above 200°C, sodium bicarbonate is formed by the reaction of solid sodium hydroxide with gaseous carbon dioxide.

CONCLUSIONS

These few examples should be sufficient to illustrate the feasibility of using equilibrium reactions in the solid state for synthetic purposes. The examples do not provide any great excitement since the products are readily available from other inexpensive processes.

The reactions in the expanded sodium carbonate lattice do afford economically viable processes. The unexpected reactivity of carbonate with carbon monoxide allows intracrystalline equilibrium processes that promise practical utility in the synthesis of a wide variety of organic compounds. The fixation of nitrogen described in this chapter, while intriguing, is likely to be far less important than other possibilities discussed in Chapter VI. The prospects of using reactions analogous to those described in Chapter IV, using only gaseous reactants and producing only volatile products are most exciting.

The chemical reactions of carbon monoxide discussed in this chapter, while unusual are not unprecedented. Transition metal carbonyl chemistry[35-37] and borane carbonyl chemistry[38] both promote carbenoid behavior of this reagent. This behavior of carbon monoxide, then, apparently arises as a consequence of some kind of electron-deficient center for the reactions involved. Both main-group elements and transition metals are effective in this role.

It is not surprising that the carbonate ion in any ionic metal carbonate is electron deficient as a consequence of its environment within the crystal lattice. One of the most electron-deficient carbons will be that in the crystal lattice of magnesium carbonate. This ionic crystal possesses such a formidable crystal lattice energy, however, that expanded lattices of this salt are not readily available. Lithium bicarbonate, however, is likely to be the best choice for decomposition to an expanded lattice of the carbonate. Methods of producing useful expanded lattices containing carbonates of cations with high charge density will be discussed in Chapter VI. It is a simple matter to prepare such an expanded lattice of sodium carbonate, as previously described. This expanded lattice allows ready access of the carbon monoxide molecule to the electron-deficient carbon of the carbonate ion.

There is an unusually high negative charge density upon the central carbon atom of carbon suboxide.[39–42] A complex between the electron-deficient carbon atom of the carbonate and the central carbon of the suboxide is quite likely. Such a complex would enhance the stability of the 1,3-biradical state required for reaction with elemental nitrogen. The chemistry described here flows naturally from this state of affairs.

REFERENCES

1. J. Chatt and G. J. Leigh, *Chem. Soc. Lond. Rev.*, 1(1), 121–124 (1972). See also: J. Chatt, *Bull. Soc. Chim. Fr.*, (2), 431–435 (1972); J. Chatt, *J. Organomet. Chem.*, 100, 17–28 (1975); J. Chatt, A. J. Pearlman, and R. L. Richards, *ibid.*, 101, C45–C47 (1975).

2. F. Haber, *Chem. Ztg.*, 34, 345–347 (1910); *Z. Elektrochem.*, 16, 244–246 (1910).

3. L. Clouet, *Ann. Chim. Phys.*, 11, 30 (1791).

4. J. J. Berzelius, *Jahresber.*, 21, 80 (1842)

5. R. Bunsen and L. Playfair, *j. prakt. Chem.*, 42, 392–400 (1847).

6. L. Thompson, *Dinglers Polytech. J.*, 23 ,281 (1839).

7. J. E. Bucher, *Ind. Eng. Chem.*, 9, 233–253 (1917).

8. E. W. Guernsey, J. Y. Yee, J. M. Braham, and M. S. Sherman, *Ind. Eng. Chem.*, 18, 243–248 (1926).

9. N. F. Murphy, R. S. Tinsley, and R. C. Hart, *Bull. Virginia Polytechnic Inst., Eng. Exp. Sta. Ser. No. 119*, 1(9), 3–9 (1957).

10. L. Farnell and L. Radom, *J. Am. Chem. Soc.*, 106, 25–28 (1984).

11. H. J. Shine and J. P. Stanley, *J. Org. Chem.*, 32, 905–910 (1967).

12. R. D Chambers, M. Clark, J. R. Maslakiewicz, W. K. R Musgrave, and K. C. Srivastava, *J. Chem. Soc. C*, 1974, 1513–1517 (1974).

13. F. Lahmani and N. Ivanoff, *Tetrahedron Lett.*, **40**, 3913–3917 (1967).
14. R. D. Chambers, J. A. H. MacBride, J. R. Maslakiewicz, and K. C. Srivastava, *J. Chem. Soc. C*, **1975**, 396–400 (1975).
15. R. D Chambers, W. K. R Musgrave, and C. R. Sargent, *J. Chem. Soc. C*, **1981** 1071–1077 (1981).
16. L. Batt, *The Chemistry of Triple-Bonded Functional Groups*. S. Patai and Z. Rappoport, Eds., John Wiley & Sons, New York, 1983, Chap. 2, Supplement C.
17. M. Levey, *Chemistry and Chemical Technology in Ancient Mesopotamia*, Elsevier Publishing Company, New York, 1959, Chap. 9.
18. E. Brouns, J. W. Visser, and P. M. de Wolff, *Acta Crystallogr.*, **17**, 614–619 (1964).
19. G. C. Dubbeldam, and P. M. de Wolff, *Acta Crystallogr.*, **B25**, 2665–2667 (1969).
20. P. M. de Wolff and W. van Aalst, *Acta Crystallogr.*, **A30**, 777–785 (1974).
21. W. van Aalst, J. den Hollander, W. J. A. M. Peterse, and P. M. de Wolff, *Acta Crystallogr.*, **B32**, 47–58 (1979).
22. C. J. de Pater, *Physica B + C*, **96**(1), 89–95 (1979).
23. J. Tcherniac, *Ger. Pat.* #145,748 (1903).
24. J. Tcherniac, *Ger. Pat.* #160,637 (1925).
25. Metzger, *U. S. Pat.* #1,439,909 (1922).
26. F. V. Bichowsky, *Ind. Eng. Chem.*, **17**, 939–940 (1925).
27. H. von Rieken, *Liebig's Annalen*, **79**, 77–79 (1851).
28. S. Misami and T. Kaneda, Chapter 16, *The Chemistry of the Carbon–Carbon Triple Bond*, S. Patai, Ed., John Wiley & Sons, New York 1978.
29. O. Diels and B. Wolf, *Ber. Dtsch. Chem. Ges.*, **39**, 689–697 (1906). See also: B. C. Banerjee, T. J. Hirt, and P. L. Walker, Jr., *Nature*, **192**, 450451 (1961).
30. L. O. Brockway and L. Pauling, *Proc. Natl. Acad. Sci., U. S. A.*, **19**, 860–867 (1933).
31. R. L. Livingston and C. N. R. Rao, *J. Am. Chem. Soc.* **81**, 285–287 (1959).
32. H. Smith and J. J. Barrett, *J. Chem. Phys.*, **51**, 1475–1479 (1969).
33. D. W. Breck, *Zeolite Molecular Sieves: Structure, Chemistry, and Use*, John Wiley & Sons, New York, 1974.
34. R. M. Barrer, Chap. 6, *Zeolites and Clay Minerals as Sorbents and Molecular Sieves*, Academic Press Inc., New York , 1978
35. W. Reppe and W. J. Schweckendiek, *Justus Liebig's Ann. Chem.*, **560**, 104–116 (1949).
36. W. Reppe and H. Vetter, *Justus Liebig's Ann. Chem.*, **582**, 133–161 (1953).
37. B. W. Howk and J. C. Sauer, *J. Am. Chem. Soc.*, **80**, 4607–4608 (1958).
38. H. C. Brown, *Organic Syntheses via Boranes*, John Wiley & Sons, New York, 1975, Chap. 7.

39. U. Gelius, C. J. Allen, D. A. Allison, H. Siegbahn, and K. Siegbahn, *Chem. Phys. Lett.*, **11**, 224 (1971).
40. J. F. Olsen and L. Burnelle, *J. Phys. Chem.*, **73**, 2298–2304 (1969).
41. J. R. Sabin and H. Kim, *J. Chem. Phys.*, **56**, 2195–2198 (1972).
42. E. A. Williams, J. D. Cargioli, and A. Ewo, *J. Chem. Soc.*, **1975**, 366–367 (1975).

Chapter VI

Reactions in the Solid State

EXPERIMENTAL BEGINNINGS

The experimental work on the subject of this chapter is in an exploratory state similar to the initial qualitative studies of the redox equilibria that led to the work reported in Chapter IV. The results obtained to date offer some encouragement.

It is ironic that the best studied reaction in the solid state is the fixation of nitrogen. It is the least studied in the case of the high-pressure redox—disproportionations. In the chemistry described in Chapter IV, graphite was chosen as the primary carbonaceous reactant for the qualitative studies. The only reactants employed in the qualitative experiments in the present chapter have been carbon monoxide and water.

In this discussion the mechanistic approach that proved effective in the earlier work is used here. Mechanistic models will be devised that are consistent with what is already known or surmised. Experiments to test these models will result in observations that require modifications of the model and so forth.

As in the earlier studies, the most difficult aspect of the experimental work is to develop procedures and techniques readily adapted to yield quantitative data. In the present studies of reactions in the solid state this facility is developing as the experimentalists' ignorance diminishes. The chemistry discussed at this point in the effort is by nature speculative. Every attempt has been made to have the mechanistic model conform to currently known chemistry. This is not a particularly limiting restriction in discussions of the reactions of carbon suboxide.[1-5] This reagent has received only moderate attention in applications to synthesis.[6]

HIGH-PRESSURE AND SOLID-STATE REACTIONS COMPARED

It is possible to draw an analogy between the ketene and polyketene anions that appear to dominate the mechanisms in the high-pressure reactions and the carbon suboxide, C_3O_2, that operates in solid sodium carbonate. The analogy suffers many defects when close parallels are drawn. It has not been possible to prepare many of the compounds that are common in the product rosters of Chapter IV. The clamped equilibrium condition that apparently controls the course of reactions occurring in solid sodium carbonate lies between reactants and products in the gaseous state. It is also likely that the nature of the voids in which the equilibrium stack must exist is much more limiting than the liquid volume of the high-pressure reactions. As a consequence, the equilibrium reactions are probably limited in number and the complexity of interactive equilibria greatly diminished.

The fixation of nitrogen by carbon in the solid state, as discussed in Chapter V, was effected at temperatures near 500°C. It has not been determined as yet that such high temperatures are required for the reaction of carbon monoxide with nitrogen. It is possible that this reaction, employing as it does only gaseous reactants, might proceed at a useful rate with considerably lower temperatures. Carbon monoxide does react with expanded sodium carbonate at 300°C to produce a dispersion of graphite. If, as is presently thought, this requires the formation of carbon suboxide, then this is a viable reactant at that temperature.

The importance of this is apparent when another difference between the high-pressure reactions and the solid state reactions is considered. It is possible to lower the temperature of the entire equilibrium system in the high-pressure apparatus and examine the products that are stable at room temperature. In the solid state reactions the products must be stable at the gas temperature in the reactor. Thermal stability at 300°C is not a greatly limiting property for most organic products arising from reaction of the malonic anhydride.

In the highly polar media in which the redox—disproportionations of Chapter IV were established, the products reflected this environment. It was rare among the many product rosters of those experiments to see difunctional molecules. No dicarboxylic acids were found, and only in those cases where adducts such as aluminum oxide were used were products arising from terminal diols found. In the reactions of carbon suboxide and its telemers dicarboxylic acids and their derivatives should be likely products.

It is possible for the system under discussion to alter the relative activities of the two prototypical reactants. Carbon monoxide that has passed through water maintained at a specific temperature will become saturated with water vapor at that temperature. Since the reaction system operates at atmospheric pressure, the total pressure of the reactants is fixed. Altering the temperature of the water adjusts the relative activity of the two reactants in a simple manner.

Until more is known about the nature of differential adsorption rates in the expanded lattice of sodium carbonate, the most sensible approach is to adjust the relative reactivities to those required by stoichiometric equations. Several such equations can be considered. The initial studies were directed toward products with moderate thermal stabilities. Those products observed in the high-pressure reactions under water-starved conditions seemed the most profitable beginning. Phenols and methylated phenols meet these requirements for the preliminary studies. These are most likely to survive distillation from the reactor in the hot sweep gas that consists primarily of carbon monoxide and water vapor.

$$14CO + 3H_2O = C_6H_5OH + 8CO_2 \qquad (1)$$

$$17CO + 4H_2O = (CH_3)C_6H_4OH + 10CO_2 \qquad (2)$$

$$20CO + 5H_2O = (CH_3)_2C_6H_3OH + 12CO_2 \qquad (3)$$

Benzene, toluene, and the xylenes also meet these requirements where the hydrogen source, water, is limiting:

$$15CO + 3H_2O = C_6H_6 + 9CO_2 \qquad (4)$$

$$18CO + 4H_2O = C_6H_5(CH_3) + 11CO_2 \qquad (5)$$

$$21CO + 5H_2O = C_6H_4(CH_3)_2 + 13CO_2 \qquad (6)$$

Furans, pyrans, and their methyl derivatives also are probable candidates for the preliminary studies. While these are not observed under water-starved conditions in the high-pressure experiments, the preference for $\alpha{-}\omega$ difunctional intermediates, previously mentioned, could favor their formation.

$$9CO + 2H_2O = C_4H_4O + 5CO_2 \qquad (7)$$

$$12CO + 3H_2O = (CH_3)C_4H_3O + 7CO_2 \qquad (8)$$

$$15CO + 4H_2O = (CH_3)_2C_4H_2O + 9CO_2 \qquad (9)$$

ROLE OF CARBON SUBOXIDE AND ITS HOMOLOGS

Equations (1) through (9) are overall reactions presenting initial reactants and final products. In Chapter V it was determined that the most probable secondary reactant is carbon suboxide, C_3O_2. This compound is sometimes called malonic anhydride. For the purposes of this discussion the name given by the discoverers of this compound will be used.[1,2]   Malonic anhydride, used hereafter, will refer to the monohydrate of carbon suboxide, $C_3H_2O_3$.

A reasonable pathway must exist between the secondary reactant, C_3O_2, and the products of the preceding equations. Carbon suboxide is reported to disproportionate to carbon dioxide and higher suboxide homologs.[3]

$$2\,O{=}C{=}C{=}C{=}O \;=\; O{=}C{-}C{=}C{-}C{=}C{=}O \;+\; CO_2 \qquad (10)$$

$$3\,O{=}C{=}C{=}C{=}O \;=\; O{=}C{-}C{=}C{-}C{=}C{-}C{=}C{=}O \;+\; 2CO_2 \qquad (11)$$

The simplest possible products in these preliminary studies are those arising from the monohydration of carbon suboxide malonic anhydrides.   Carbon suboxide, C_3O_2, is the simplest ketene analog in the solid-state chemistry.   In the high-pressure reactions, ketene anion, $HC{\equiv}CO^-$, filled this role.   In those reactions acetate ion was the simplest monohydrate possible.   In the solid-state reactions malonic anhydride, $C_3H_2O_3$, should be equally ubiquitous.   Further reduction of this to simpler compounds also is quite probable.   Some of these will be discussed later.

$$4CO \;+\; H_2O \;=\; C_3H_2O_3 \;+\; CO_2 \qquad (12)$$

The methyl derivatives are also possible products.   Higher homologs of the suboxide will be required in order to form these more complex structures. Hydration to the polyenol acids must be rapid for all homologs of the suboxide series:

$$O\!=\!C\!=\!C\!=\!C\!=\!C\!=\!C\!=\!O + 3H_2O = HO-\overset{H}{\underset{}{C}}\!=\!\overset{H}{\underset{OH}{C}}-\overset{}{\underset{OH}{C}}\!=\!\overset{H}{\underset{}{C}}-\overset{OH}{C}\!=\!O \qquad (13)$$

$$O\!=\!C\!=\!C\!=\!C\!=\!C\!=\!C\!=\!C\!=\!O + 4H_2O = HO-\overset{H}{\underset{}{C}}\!=\!\overset{H}{\underset{OH}{C}}-\overset{H}{\underset{OH}{C}}\!=\!\overset{H}{\underset{OH}{C}}-\overset{OH}{C}\!=\!O \qquad (14)$$

The relationship between the polyhydroxyacid products of equations (13) and (14) and carbohydrates is illustrated in equation (15). This is a simple decarboxylation reaction that should occur readily:

$$HO-\overset{H}{\underset{}{C}}\!=\!\overset{H}{\underset{OH}{C}}-\overset{H}{\underset{OH}{C}}\!=\!\overset{H}{\underset{OH}{C}}-\overset{OH}{C}\!=\!O = \text{(lactone)} + H_2O + CO_2 \qquad (15)$$

The dehydrated lactone product of equation (15) possesses a sufficient number of carbon atoms to generate products containing six carbon atoms. The formation of substituted furans, pyrans, and unsubstituted phenols is likely from the product lactone of equation (15). Hydrolysis of the lactone would have to occur to allow the rearrangements required for the following two equations:

$$7CO + 2H_2O = (CH_3)C_3HO_3 + 3CO_2 \qquad (16)$$

$$10CO + 3H_2O = (CH_3)_2C_3O_3 + 5CO_2 \qquad (17)$$

The anhydrides are likely to be swept into the condenser by the sweep gases.   Water, present in the sweep gas, will be condensed along with the anhydride.   Hydrolysis to malonic acid is most probable when the trap is warmed.

The polymerization of carbon suboxide to telemeric homologs is a matter of some disagreement.[7,8]   A recent study of the water-induced polymerization describes it as a reaction between malonic anhydride and carbon suboxide with the formation of carbon dioxide and poly-α-pyrone.[9]   Apparently, many 1,3-diketones react readily with carbon suboxide in a similar fashion.[10]

For the purpose of the current reaction model this disagreement is not relevant.   The sequence of the reactions of carbon suboxide with water just presented was arbitrarily chosen from among many possible sequences.   The water-induced polymerization is a different sequence that would serve equally well. Other sequences involving the reduction of the anhydride to a 1,3-diketone would serve as well.

Such reductions can be effected readily with the reactants carbon monoxide and water.   The water-gas shift equilibrium,

$$CO \ + \ H_2O \ = \ H_2 \ + \ CO_2 \qquad (18)$$

will furnish hydrogen for hydrogenation or reduction of any molecule that reacts more avidly with hydrogen than does carbon dioxide.   Another reaction involving carbon suboxide is perhaps more interesting in considering the analogy between the reactions in the high-pressure chemistry and the present case.   Under conditions of low activity of the suboxide, any kind of polymerization is not favored.   The equilibrium of the following equation is favored under such conditions:

$$O=C=C=C=O \ + \ H_2O \ = \ H_2C=C=O \ + \ CO_2 \qquad (19)$$

The equilibrium of equation (20) may be the sum of two equilibria. The carbon suboxide first hydrolyzes to malonic anhydride and then pyrolyzes to the ketene.[11] Substituted ketenes also pyrolyze in this fashion.[12,13]

$$C_3H_2O_3 \ = \ H_2C=C=O \ + \ CO_2 \qquad (20)$$

Under the conditions of these reactions in an expanded lattice of sodium carbonate, acetic anhydride is to be expected as the volatile product arising from the low activity of carbon suboxide:

$$2O=C=C=C=O \ + \ 3H_2O \ = \ (CH_3CO)_2O \ + \ 2CO_2 \qquad (21)$$

The overall reaction to form acetic anhydride from the proto-typical reactants, carbon monoxide and water, is:

$$8CO \ + \ 3H_2O \ = \ (CH_3CO)_2O \ + \ 4CO_2 \qquad (22)$$

This reaction is seen to require a stoichiometry with the lowest carbon-monoxide-to-water ratio of any example that results in reduced compounds of carbon. This is the reaction that will require the highest water activity to be favored. Even higher water activities are required to favor the production of hydrogen gas as a product.

The reaction also provides a nonpolar analog of the acetate ion found in most of the reactions under autogenous pressure discussed in Chapter IV.   This could be of great importance when this study of intracrystalline equilibria proceeds to the point where mixed lattices of the carbonate with oxide adducts such as boric oxide and aluminum oxide are employed.

All of the possibilities listed in the preceding equations are encompassed by carbon-monoxide-to-water ratios lying between 6 to 1 and 2 to 1.   It is not likely that adjustment of reactant ratio will allow a very precise selection of favored product.   In general, however, high ratios (low water activity) should favor aromatic compounds and low ratios (high water activity) will favor the more highly hydrogenated products such as acetic anhydride.   In the reactions under autogenous pressure this was seen to be a consequence of the role of water as the sole source of hydrogen for the disproportionation reactions.   The reasons are the same here.   The water-gas shift reaction and its analog in solution is a ready rationale for this relationship.

EXPERIMENTAL PROCEDURES AND RESULTS

The solid-state reactions occur in the expanded lattice of sodium carbonate and should be quite susceptible to changes in pore size.   This is a parameter that can be adjusted incrementally as has been previously discussed.   Selectivity in these reactions may depend upon this parameter more than upon the various activities of reactants.

The standard form of sodium carbonate used in the preliminary studies is that obtained from the pyrolysis of pure sodium bicarbonate in a stream of argon at 300°C. A mixture of carbon monoxide and water vapor is prepared by injecting water through a hypodermic syringe needle into the stream of carbon monoxide as it enters the heat exchanger. The water was injected by means of an adjustable metering pump. Different rates of water injection relative to the carbon monoxide flow rate produces different reactant compositions. For most of the screening reactions the reactant mixture is estimated at 80 mol% of carbon monoxide. This mixture was heated to a temperature that produced a pellet temperature of 400°C in the reactor. These conditions apparently avoided a buildup of elemental carbon in the carbonate lattice since the pellet remained white throughout the experiment.

The percent conversion for each experiment was estimated by comparing the carbon dioxide produced in the reaction, condensed in the -196°C trap, with the amount of carbon dioxide produced by burning the unchanged carbon monoxide exiting the reactor. The amount of carbon dioxide produced in the reactor is a measure of the amount of redox–disproportionation that occurs during the entire period of reaction.

Most products are condensed along with water in the -78.6°C trap. This mixture is extracted with ether and the resulting ether solution concentrated on the vacuum apparatus. The oil produced by this procedure is analyzed by gas chromatography mass spectrometry. Alternatively, the ether extract is introduced to the vacuum apparatus and subjected to fractional condensation. The different fractions obtained from this procedure are examined and identified, where possible, by spectrometry.

In the experiments to date, furans, pyrans, phenols, and other aromatics, such as benzene and toluene have been detected among the products of this reaction. Hydrogen and methane in varying amounts are often found.

No acetic acid has been observed, although any acetic anhydride formed in the reaction and condensed along with water in the $-78°C$ trap should hydrolyze readily. The reactant mixture is 80 mol% of carbon monoxide, well above the 66 mol% of carbon monoxide required by the stoichiometry of the acetic anhydride equilibrium described by equation (22). The lack of detection of acetic anhydride among the products of these reactions may also indicate that adsorption processes may require an even higher activity of water than indicated by simple stoichiometry.

There also is the possibility that carbonate in the crystal lattice is being replaced by acetate. Reactant water may be at a sufficiently high activity in the voids to hydrolyze any acetic anhydride to acetic acid. The acetic acid would replace carbonate with the formation of carbon dioxide. Changes in pellet appearance should accompany this chemistry if the runs were continued for sufficiently long periods. Freeing acetate from the crystal lattice may require adjusting the reactant stream to include a steady-state concentration of carbon dioxide higher than that produced in the redox reaction.

Recycling the reaction mixture to a steady-state condition is a more attractive alternative, but this procedure will require major modification of the reaction system. More general knowledge of the important reactions in this chemistry will permit such modifications to be made in a sensible fashion.

The reactions have been standardized on runs of 4 hr. This may not be sufficiently long to produce amounts of product mixtures that partition well in the simple ether extraction of these qualitative experiments.   There is a considerable amount of carbon dioxide formed during this same period without apparent deposition of elemental carbon. This must be accompanied by an equivalence of reduction of carbon monoxide, with reduced products appearing in similar stoichiometric quantities. Conversions of up to 10% per pass have been achieved. In a recycling system these would be satisfactory. In a single-pass system they are marginal.

There may be reduced products still trapped as clathrates in the solid. The pellets will have to be dissolved and analyzed in order for this possibility to be evaluated. An expansion of the lattice could allow such entrapped molecules to equilibrate. In this case there will have to be experiments performed on the pyrolysis of hydrates of sodium bicarbonate to see if it is possible to prepare pellets of such lattices.

The exploratory reactions all have been effected at $400°C$. This temperature may be too high for optimum results as far as some expected products are concerned. Some of the expected products may participate in equilibria favored at this temperature and be removed before reaching the product analysis train. It is necessary to study the temperature dependence of the product roster before this possibility can be eliminated.

OTHER SOURCES OF HYDROGEN

A logical extension of this experimental work is to study other compounds that serve as hydrogen sources. Ammonia comes to mind as the simplest substitution. Pyrroles, pyridines, imines, and nitriles are the most likely products of binary mixtures of carbon monoxide and ammonia. Imidazoles, pyrazines, and pyridazines also are probable products. Ammonia reacts readily with carbon monoxide:

$$CO \; + \; NH_3 \; = \; HCN \; + \; H_2O \tag{23}$$

This reaction provides an opportunity to study a simple quaternary reaction mixture. In addition to the reactions already discussed, reactions of the various intermediates with ammonia and hydrogen cyanide may be invoked. A few examples should suffice to illustrate the interesting possibilities of this chemistry.

Carbon suboxide reacts rapidly with ammonia and should react similarly with hydrogen cyanide. Ammonia reportedly forms malonamide, while in an analogous reaction, aniline forms the corresponding anilide.[1] There is no report by these authors or later workers of formation of the cyclic imide. This is likely to be a consequence of a more probable pyrolysis route for the amide to malononitrile.

The reaction of the suboxide with alcohols evidently occurs in discrete steps[14] with the rate constant for the first step at least double that of the second. It is also possible to pyrolyze malonyl chloride to a ketene carbonyl chloride.[15] The first addition of hydrogen chloride to carbon suboxide is apparently similarly favored:

$$O=C=C=C=O \ + \ HCl \ = \ ClCO-CH=C=O \qquad (24)$$

$$ClCO-CH=C=O \ + \ HCl \ = \ ClCO-CH_2-COCl \qquad (25)$$

The reaction with hydrogen cyanide may be complex with typical addition analogous with that of hydrogen chloride unlikely. A more typical addition of hydrogen cyanide would be to form the cyanohydrin and the dicyanohydrin:

$$O=C=C=C=O \ + \ HCN \ = \ NCC(OH)=C=C=O \qquad (26)$$

$$O=C=C=C=O \ + \ 2HCN \ = \ NCC(OH)=C=C(OH)CN \qquad (27)$$

The products of equations (26) and (27) offer many synthetic possibilities.   They both should be unusually acidic and react avidly with water.   The cyanohydrin should readily form a lactam that is reducible to pyrrole.   The dicyanohydrin should form pyridine or substituted pyrroles in a similar series of reactions.

$$NCC(OH)=C=C=O \ + \ H_2O \ = \ \overline{HNC=C(OH)-CH(OH)-CO} \qquad (28)$$

Reactions of the cyanohydrins with ammonia could furnish intermediates to pyrazoles, pyrazines, and piperazines by analogous chemistry.   Pentacarbon dioxide and heptacarbon dioxide, or any sequence of hydration and polymerization of carbon suboxide to five- and seven-carbon structures, will produce yet more complex molecules.

There are other sources of hydrogen not capable of such direct generation of water and less likely to result in such complex reactions. It will be useful to examine acetylene in this role, since many reactions of this reactant with carbon monoxide, catalyzed by electron deficient centers, have already been reported.[16-20] Quinones, hydroquinones, and oxofurans are the products to be expected of catalysis by electron-deficient centers.

Alkanes and alkenes similarly could be used. In short, all of the hydrocarbons that have been employed as reactants in interesting syntheses catalyzed by boron compounds and transition metals or their complexes could yield similar syntheses from reactions in the solid state.

OTHER MODIFICATIONS OF THE
SOLID-STATE REACTION

Alcohols such as methanol also could function as reactants in these solid-state reactions to yield interesting and useful products. Differences in the reactivity of alcohols, compared with that of water, with carbon suboxide should produce interesting alterations in the product rosters.

It does appear that lithium bicarbonate and mixtures of the bicarbonate and carbonate pyrolyze to an expanded lattice in a fashion similar to the sodium congeners. The high charge density of the positive ion in these lattices will, no doubt, enhance the electron deficiency of the carbon in these lattices. They should function more effectively in many of the reactions. Experimental work on these salts is deferred until a better understanding of the diverse reactions in sodium carbonate is available.

The experimental work is not sufficiently advanced to begin study of modifications of the carbonate lattice by addition of other oxides. This does promise to be an area worthy of exploitation in efforts to alter the dominant reactions of the typical reaction stack. There are several ways in which the addition of boric oxide or aluminum oxide to the sodium carbonate lattice might change the reaction parameters.

Lattices of different compositions should result from the pyrolysis of mixtures of sodium bicarbonate and boric acid or hydrated aluminum oxide. This experiment is essentially an application of preparative mineralogy. It will be necessary to study a range of compositions in order to prepare useful expanded lattices. These compositions can range from a prevalent structure of sodium carbonate to a zeolite structure with sodium carbonate. As in the case of pure sodium carbonate, the density of the pyrolyzed pellet compared with that of the compressed mixture will measure the free volume in the lattice.

A modification of the same procedure will be applied to mixtures of the group III oxides with the bicarbonates of the alkaline earth metals. Expanded lattices containing these more highly charged ions should contain more electron-deficient carbon than sodium carbonate. This may not be particularly useful in these mixed salts, however, since the boron or aluminum atoms introduced into these lattices will be yet more electron deficient than the carbon of these carbonates.

Modifications containing manganese are of interest because of the specificity of the redox disproportionation under autogenous pressure, summarized in Table 4.10. Methanol and 2-butene comprised the limited roster of products for this example. In equations (31) and (33) of Chapter IV these products were rationalized. Acidification of the highly alkaline medium of the high-pressure reactions was necessary before these products could be observed.

The preparation of an expanded-lattice sodium carbonate with a manganate adduct is logically to be expected from the pyrolysis of a mixture of sodium bicarbonate and active manganese dioxide. Whether the manganate will be reduced to manganite in the lattice by moist carbon monoxide remains a question. Cyclic ethers of the manganite should be possible, from what is already known about this chemistry in the solid state. These could decompose in the lattice with the production of 2-butene, but this will only be determined by experiment. It seems improbable that any methanol can be formed in this kind of reduction.

ALUMINUM-MODIFIED REACTIONS

Acetic acid is not a primary product in the solid-state reactions. Acetic anhydride is a secondary product which should serve this role in the solid-state reactions. It may not be necessary to form the acid anhydride since ketene will react directly with any hydroxyaluminate ions in the lattice. The aluminum acetates that result can undergo the novel aluminum alkyl chemistry described in equations (81) through (88) of Chapter IV.

In addition to the aluminum alkyl chemistry and the synthesis of secondary and tertiary carbinols, tetrahydrofuran is also probable from these modifications. The solid-state reactions appear to be a superior method for obtaining furans, and the addition of aluminum oxide could enhance these syntheses by intercepting an intermediate prior to the formation of carbon suboxide. The pertinent reactions are given in equations (29) through (31).

$$
{}^-O\text{-}\underset{\overset{\|}{O}}{C}\text{-}\underset{\overset{\|}{O}}{C}\text{-}O^- \; + \; 2CO \;=\; {}^-OC\text{-}\underset{\overset{\|}{O}}{C}\text{-}\underset{\overset{\|}{O}}{C}\text{-}CO^- \tag{29}
$$

$$
{}^-OC\text{-}\underset{\overset{\|}{O}}{C}\text{-}\underset{\overset{\|}{O}}{C}\text{-}CO^- \; + \; {}^-OAl(OH)_3 \;=\; (O^-)_2Al\underset{\diagdown O\text{-}\underset{\overset{\|}{O}}{C}\text{-}C=O}{\overset{\diagup O\text{-}\underset{\overset{\|}{O}}{C}\text{-}C=O}{}} \; + \; OH^- \; + \; H_2O \tag{30}
$$

Reduction of the cyclic product of equation (30) should occur readily in the presence of the carbon monoxide–water reactant mixture. Such reductions have been noted in several reactions in the expanded lattice sodium carbonate.

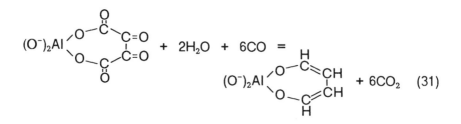

Hydrolysis to furan will complete the catalytic cycle:

$$(O^-)_2Al \overset{O-\overset{H}{C}\,\overset{}{\diagdown}CH}{\underset{O-\underset{H}{C}\,\diagdown CH}{}} + H_2O = (O^-)_2Al(OH)_2 + O \overset{\overset{H}{C}\,\diagdown CH}{\underset{\underset{H}{C}\,\diagdown CH}{}} \quad (32)$$

This intervention and the sequence of reactions (30) through (32) could preclude the formation of significant amounts of carbon suboxide. The many equilibria utilizing this secondary reactant then could not function at a practical level. As a consequence, simpler product rosters may well be the rule for aluminum-modified reactions.

CATALYTIC USES OF ZEOLITES

There are several examples where the catalytic activity of zeolites is exploited. In most of the cases where the zeolites serve as effective catalysts the product rosters reported are relatively simple. This may indicate that an intervention by aluminate, such as that just discussed, operates to preclude formation of carbon suboxide. Carbon suboxide apparently acts to produce more different products than is typical of the zeolite chemistry.

Zeolites illustrate the utility of chemically active molecular sieves. There seems to have been little interest in creating a reaction model for these interesting catalysts. In general, studies have compared different zeolites using the critical parameter of pore size. Differences in the chemical action of individual zeolites could be due to this parameter or to other chemical factors that accompany the procedures for producing the different pore sizes.

Much of this catalysis work converts synthesis gas: a mixture of carbon monoxide and hydrogen in a mole ratio of 1 to 2. The synthesis gas can be prepared by the reaction of water with elemental carbon. Initial reaction of the coke and water is promoted at $1000°C$. This produces water gas, an equimolar mixture of carbon monoxide and hydrogen:

$$C + H_2O = CO + H_2 \tag{33}$$

The water gas is cooled to approximately $700°C$ and additional water injected to oxidize the carbon monoxide to carbon dioxide. This is the water-gas shift reaction:

$$CO + H_2 + H_2O = CO_2 + 2H_2 \tag{34}$$

Equal volumes of the product mixtures of equations (33) and (34) are combined. The carbon dioxide is removed from this mixture, usually by Joule–Thomson effect condensation. Synthesis gas is prepared to this composition primarily as a consequence of the first large industrial use of this mixture, catalytic synthesis of methanol. Zeolites were used as catalysts for this purpose early in the development of this large industrial process.

Earlier, methanol and many other alcohols and oxygenated solvents were prepared using processes based upon the reactions of acetylene with air and water. Much of this industrial acetylene chemistry was located in the Ohio River valley near large population centers. The amount of acetylene gas required to keep these many plants functioning full time was intimidating. Well-justified fears that acetylene's unpredictable behavior could cause a civil disaster of monumental proportions led to the abandonment of acetylene chemistry in the American chemical industry. Many among the readers may remember the brief period of acute shortages of ethylene glycol antifreeze that accompanied the abandonment of industrial-scale acetylene chemistry. Substitution of alternate processes rapidly brought these shortages to a manageable level. While zeolite-catalyzed chemistry is not presently as versatile as the acetylene chemistry it has displaced, it is considerably safer to execute.

The most useful zeolites have proved to be those low in aluminum content, with the majority of the lattice consisting of silica. The most versatile group apparently consists of assemblies of five- and ten-membered rings fused to an open structure of high thermal stability.[21] Varying the aluminum content can greatly alter the nature of the catalyst.[22] In general the activation of the aluminum site by dehydration to enhance its Lewis or Brønsted acidity is suggested. The implicit model suggests that carbonium ion chemistry is dominant. There has not been any mechanism put forward that suggests chemical reduction of the aluminum to produce an electron-deficient center capable of promoting chemistry similar to borane catalysis or transition metal catalysis.

The chemical significance of electron-deficient centers in the lattice apparently has not been investigated in a systematic fashion. This is understandable since aluminate centers probably dominate such a role in zeolites. Mixed aluminate–borate lattices, which could afford some control over the variable of interest, are currently under investigation.[23] There are catalytic enhancements, but these have not been evaluated.

Haber's nitrogen fixation[24] may well be an example where zeolite structures, modified by the introduction of novel electron-deficient iron centers, effect useful chemistry. Catalyst preparation for this process generally involves procedures that result in expanded structures.

New Zealand's large scale gasoline plant, operating as it does with a methane-based, synthesis-gas feed stock, emulates Fischer–Tropsch chemistry.[25] Similarly, the catalytic reforming capabilities of zeolites would appear to mimic transition metal chemistry.[26] The zeolite catalyst in these processes also, undoubtedly, demonstrates the electron-deficient nature of such catalysts.

The present work will allow a wide range of compositions where the aluminate or borate activity in the lattice can be varied at will. This can be accomplished without greatly altering pore sizes since the carbonate–bicarbonate ratios can be varied over such a wide range. It will be possible in this study to assess the effect of controllably increased activity of the aluminate—borate electron deficient centers. The knowledge gained from the study of the expanded-lattice sodium carbonate furnishes a comparison against which the role of these electron-deficient centers can be assessed in a systematic way.

SUMMARY

The reactions in expanded lattices of ionic salts apparently parallel the reactions under autogenous pressure in closed vessels. There are some differences observed and expected between the two classes of reaction, however. The reactions discussed in Chapter IV were multiphase equilibria involving solid phases, a liquid phase, and a gas phase. Those discussed in chapters V and VI apparently involve only the gas phase and the solid phase, unless the adsorbed phase functions as a separate phase analogous to the liquid phase.

There are additional ways of utilizing the various equilibria operating in both of these systems. The aluminum alkyl chemistry carried out in bench top experiments, mentioned in Chapter V, involves a set of equilibria that are not pressure dependent. A more exact duplication of many of the reactions described in Chapter IV would be reactions in fused potassium hydroxide with a water solute at temperatures where the water vapor pressure above these solutions is less than 1 atm. For potassium hydroxide with 15% water this temperature would be in the vicinity of $225^{\circ}C$. Many of the high-pressure reactions proceed at this lower temperature. As with most such reactions, these are still virtually instantaneous equilibria. A second phase, a liquid high-molecular-weight alkane or alkyl aromatic compound that does not react with molten potassium hydroxide at these temperatures, would create a multiphase analog of the high-pressure reaction. Distillation of volatile products from this oil phase could maintain an equilibrium clamp dependent upon the volatility of the product and the partition of the product between the potassium hydroxide phase and the organic layer. Changing the parameters of this clamp could afford a more selective control of the product roster.

This flight of fancy, as well as the many defects remaining in our understanding of the elementary equilibrium chemistry of carbon, is a persuasive argument that an enormous backlog of experiments exists, waiting to be attempted. There is little probability that future generations of experimental chemists and engineers will lack useful employment.

REFERENCES

1. O. Diels and B. Wolf, *Ber. Dtsch. Chem. Ges.*, **39**, 689–697 (1906).
2. O. Diels and P. Blumberg, *Ber. Dtsch. Chem. Ges.*, **41**, 82–86 (1908)
3. A. Klemenc and G. Wagner, *Monatsh. Chem.*, **66**, 337 (1935).
4. A. Klemenc and G. Wagner, *Ber. Dtsch. Chem. Ges.*, **70**, 1880–1882 (1937).
5. L. H. Ryerson and K. Kobe, *Chem. Rev.*, **7**, 479–492 (1930).
6. T. Kappe and E. Ziegler, *Angew. Chem. Int. Ed. Engl.*, **13**, 491–504 (1974).
7. O. Diels, *Ber. Dtsch. Chem. Ges.*, **71**, 1197–1200 (1938).
8. A. Klemenc, *Ber. Dtsch. Chem. Ges.*, **71**, 1625–1626 (1938).
9. H. Sterk, P. Tritthart, and E. Ziegler, *Monatsh. Chem.*, **101**, 1851–1854 (1970).
10. A. Omori, N. Sonoda, and S. Tsutsumi, *J. Org. Chem*, **34**, 2480–2482 (1969).

11.   O. Diels and G. Meyerheim, *Ber. Dtsch. Chem. Ges.*, **40**, 355–363 (1907).

12.   H. Staudinger and E. Ott, *Ber. Dtsch. Chem. Ges.*, **41**, 2208–2217 (1908).

13.   P. Lorenčak, J. C. Pommelet, J. Chuche, and C. Wentrup, *Chem. Comm.*, **1986**, 369–370.

14.   H. Binder and W. Lindner, *J. Chromatogr.*, **77**, 323–329 (1973).

15.   E. Ziegler and H. Sterk, *Monatsh. Chem.*, **98**, 1104–1107 (1967)

16.   W. Reppe and W. J. Schweckendiek, *Justus Liebig's Ann. Chem.*, **560**, 104–116 (1949).

17.   W. Reppe and H. Vetter, *Justus Liebig's Ann. Chem.*, **582**, 133–161 (1953).

18.   B. W. Howk and J. C. Sauer, *J. Am. Chem. Soc.*, **80**, 4607–4608 (1958).

19.   N. E. Schore, B. E. La Belle, M. J. Knudsen, H. Hope, and X. J. Xu, *J. Organomet. Chem.*, **272**, 435–446 (1984).

20.   W. Hubel, *Organic Synthesis via Metal Carbonyls*, vol. 1, 273, I. Wender and P. Pino, Eds, Interscience, New York, 1969.

21.   P. B. Weisz, *Chemtech*, **1987**, 368–373.

22.   G. T. Kokotailo, S. L. Lawton, D. H. Olsen, and W. M. Meier, *J. Phys. Chem.*, **85**, 2238–2243 (1978).

23.   E. G. Douraine et al., in *Catalysis by Acids and Bases*, B. Imelik, Ed., Elsevier, Amsterdam, 1985; p. 135.

24.   F. Haber, *Chem. Ztg.*, **34**, 345–347 (1910); *Zeit. Elektrochem.*, **16**, 244–246 (1910).

25.   C. D. Chang and A. J. Silvestri, *Chemtech*, **1987**, 624–631 (1987)

26.   S. L. Meisel, *Chemtech*, **1988**, 32–37 (1988).

Subject Index

Acetaldehyde, from pyrolysis of tribochemical reaction products, 55
Acetate:
from,
  acetylene/alkali reaction, 81
  carbonaceous reactant/OH⁻ reactions, 88–89
  carbon monoxide/hydroxide reaction, 140
  graphite/hydroxide reaction, 26
  methane/OH⁻/carbonate reaction, 91
  Varrentrap reaction, 110
generation of methane from, 72, 91
oxidation to methanol, 93, 122
reaction with aluminum oxide, 135–136
replacement of carbonate in crystal lattice by, 194
Acetic acid:
chemistry of, 70–76
dehydration of, 70
from,
  carbon/water reaction, 82
  cellulose/OH⁻ reaction, 105
  charcoal production, 13
  formaldehyde/water reaction, 81
  ketene and water reaction, 69
  tribochemical reaction of graphite/ water, 55
Acetic anhydride:
as secondary product in solid state reactions, 200–201
from carbon monoxide/water in sodium carbonate lattice, 191
hydrolysis product from carbon suboxide, 191
Acetone:
catalytic activity with active graphite, 47

from,
  acetate/methylide, 73–74
  alkali salts of acetic acid, 71, 71–72
  calcium acetate, 71
  charcoal production, 13
  sodium orthoacetate, 71
Acetylene(s):
aromatic acetylene reactions, 8
as interstellar molecules, 2
formation of aldehydes from, 4
from,
  acetylide/water reaction, 81
  carbonaceous chondrites, 7
  charcoal combustion, 13
  pyrolysis of tribolyzed graphite, 53
reaction with,
  fulvenoid graphite, 43–47
  molten alkalies, 81
Acetylide:
formation of, 80
reaction with water, 80, 81
Acrylate:
from ketene/formate, 143–144
reduction to propionate, 144
Alanine, mechanism of formation in N_2/ graphite/molten KOH, 154–155
Alcohols:
aliphatic alcohols from charcoal, 13
from,
  Al_2O_3 or B_2O_3/carbonaceous reactant/OH⁻ reactions, 89
  cellulose/calcium hydroxide equilibrium, 115
  disproportionations with Al_2O_3, 130–131
  hydrolyses of aluminum alkoxides, 131